THE REBIRTH OF COSMOLOGY

Translated from the French by Helen Weaver

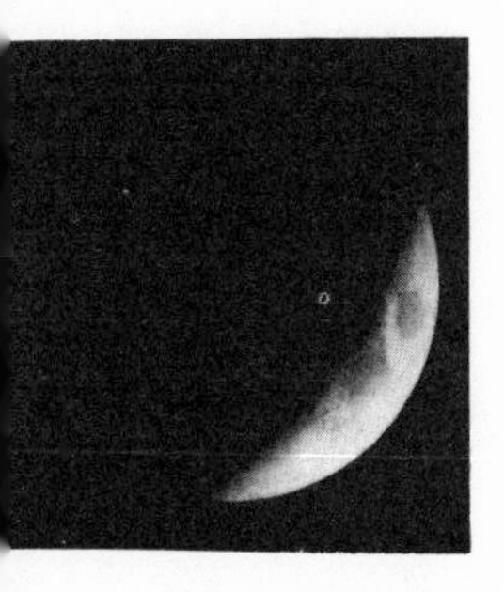

THE REBIRTH OF COSMOLOGY

By Jacques Merleau-Ponty and Bruno Morando

Alfred A. Knopf ▪ New York ▪ 1976

 Published in the United States by Alfred A. Knopf, Inc., New York, and simultaneously in Canada by Random House of Canada Limited, Toronto. Distributed by Random House, Inc., New York.
Originally published in France as
Les trois étapes de la cosmologie by Éditions Robert Laffont, S.A., Paris.

Library of Congress Cataloging in Publication Data

Merleau-Ponty, Jacques.
The rebirth of cosmology.

Translation of Les trois étapes de la cosmologie.
Bibliography: p.
Includes index.
1. Cosmology. I. Morando, Bruno, joint author.
II. Title.
QB981.M4713 1976 523.1 74-21319
ISBN 0-394-48267-0

Manufactured in the United States of America
First American Edition

CONTENTS

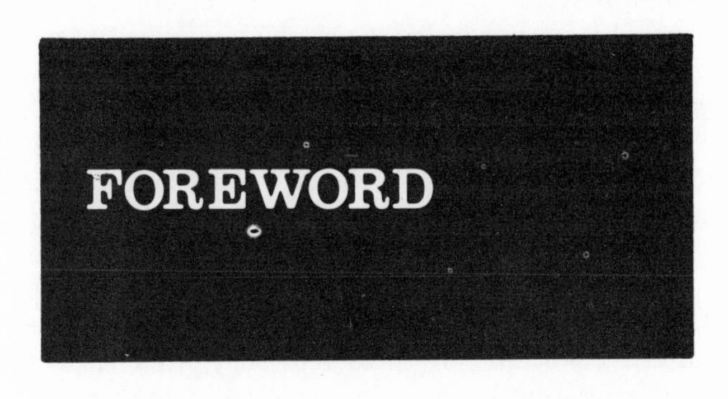

This book is the result of a collaboration between two writers, one an astronomer, the other a philosopher. An association of this kind is in no way unusual. What is less common is the particular form adopted by this collaboration: the book consists of two essays written independently and presented in parallel fashion, according to a single overall plan. The nature of the subject dictated not only an account that was both epistemological and historical, or philosophical, if you will, but also an account that was strictly scientific. Also, each of the authors wanted to be completely unhampered by the concerns of the other; one wanted to be able to write without equations, while the other wanted to give his discussion a clarity and precision that would inevitably be blurred by considerations of genesis and evolution, inquiries into the how and the why, especially in a science as exceptional as cosmology.

For this reason a system of rather numerous and detailed references has been provided so that the reader can, if he chooses, refer to the scientific facts and theories to which the philosophical account alludes. It did not seem necessary to provide a reverse system. The reader will choose his own way of reading the book in accordance with his background and personal interests; the authors have no particular advice to give him in this regard except on the method of understanding and utilizing the references.

The book is divided into three parts, which are independent in the sense that in each part, the philosophical account and the scientific account are presented separately. References such as "(IIb3)" appearing in the philosophical text must therefore be understood as referring to the corresponding divisions of the scientific account *in the part that one is reading*. In exceptional cases, the author of the philosophical discussion has been led to refer to the scientific discussion in another part. In such cases, this is explicitly mentioned in the reference. For example, if in the philosophical discussion of Part Two, allusion is made to scientific data furnished in Part Three, the reference will take the following form: (Part Three: Ic4).

INTRODUCTION

The book we are presenting to the reader is not an introduction to astronomy or astrophysics. Such works are not rare and there are some very good ones; from year to year, as advances are made in science, new works are published by which the public can keep abreast of the latest developments. The book you are about to read is something else, something less usual and more adventurous; what we are presenting here is an introduction to cosmology. The difference is probably not difficult to understand; nevertheless, it deserves some comment.

Astronomy observes the stars and systematically describes their properties, their groupings, their motions, both apparent and real. Astrophysics tries to interpret these properties in terms of known physical laws, which it applies to more or less simplified models of the systems being observed. These are *sciences* which have in common with the other sciences something which differentiates them from cosmology. To understand this difference, let us consider another and very different science, biology. Here the object and methods are quite different. The stars are enormous objects, they are extremely far away, their evolution is very slow by the scale of ordinary terrestrial phenomena, and all experimentation on them is ruled out by the inadequacy of our means. On the other hand, their structure is relatively simple, so that the fundamental laws of physics may be applied to them rather directly. Living creatures, on the contrary, are near at hand and available for all kinds of experiments; yet their structure is almost hopelessly complex and their behavior seems to defy any reduction to the elementary laws of the physical world. Despite these major differences, there are some essential analogies. In both astrophysics and biology, there occurs some sort of preliminary selection of a certain area of experience which, although its boundaries are certainly not fixed in any absolute or invariable way, definitively *specializes* the research of the biologist in a direction that is different from the research of the astronomer.

It is in precisely this way that cosmology differs from all the other sciences, and it is for this reason that it should not be confused *a priori* with astronomy any more than with biology. For the proper object of its study is not the stars, living creatures, or inanimate matter, or the carbon compounds, or human societies; the proper object of its study is the universe considered *as a whole*. This does not mean that cosmology includes all the other sciences, that it is an encyclopedia; but it does mean that all the sciences have or may have some relation to it, insofar as the objects of their study have some relation to the totality.

Plato and Aristotle, who were the founders of cosmology, as of many other things, believed that the stars and the universe were *alive;* it is clear that, according to such a conception of cosmology, astronomy and biology are not really separable. This idea, one suspects, is altogether excluded from modern cosmological theories, according to which life is not a property that can reasonably be attributed either to the universe, considered as a whole, or to any of the great systems such as stars, nebulae, planets, star groups, galaxies, etc., whose structure and evolution astrophysics reconstructs. Life seems to be limited to objects of insignificant dimensions, "minuscule creatures that crawl on the surface of the stars,"* creatures confined to at least one planet during part of their existence (to others, perhaps, but up to the present we know of none). But *this very fact*, namely, the evident rarity of life in the universe and the fact that our existence appears not to be associated with part of a living creature greater than ourselves (as our red corpuscles are associated with ourselves), these very statements which have to do not with life, but with the relation of life to the whole, belong not to biology but to cosmology, because they strip the universe of a property which—as we see from the mistake of the ancients—was spontaneously attributed to it.

Distinct from all the sciences, but potentially associated with all of them, is cosmology itself a science? A question of this kind does not admit of a simple or easily verifiable answer; and the whole of the book you are about to read may not be enough to give a satisfactory reply.

In the conception of rational knowledge that prevailed in Europe until the end of the seventeenth century, cosmology (at that time imperfectly distinguished from physics), although not the highest science, was at least an essential element of that supreme system of thought called philosophy. But in the seventeenth century a new science—Newton was still calling it "natural *philosophy*"—which was gradually to become what we call physics, altered the position of cosmology in the system of sciences to such a degree that cosmology seemed destined to disappear gradually into the mists of the unknowable.

Nothing could be more significant than this decline of cosmology in the age that may be called classical when one is speaking of science (between the end of the seventeenth century and the end of the nineteenth century). And nothing could be more significant than the meditation on cosmology's decline of a very great philosopher, Immanuel Kant. To Kant we owe one of the first coherent attempts at cosmological synthesis, which is based on Newton's law of universal gravitation, and which

* Henri Poincaré, *Leçons sur les hypothèses cosmogoniques*, preface.

on certain points anticipates the theories of the twentieth century to a remarkable extent. This was in the middle of the eighteenth century; but thirty years later, when Kant wrote the work that assured him immortality, he claimed to prove that the ultimate questions that cosmology might raise—whether the universe is finite in space, whether it has an origin in time—are insoluble because they have no meaning, and that in trying to answer them one can only become ensnared in contradiction. Thus, in *The Critique of Pure Reason* he was forced, as the century and his own thinking advanced, to renounce cosmology and to renounce it for human reason once and for all. Whether or not Kant was right, this renunciation became widespread among the scientists of the nineteenth century, without, of course, being universal. Yet it is clear that, in fact, this alleged renunciation was accompanied by a belief in hypotheses which, beneath their simple and obvious appearance, were in reality of cosmological significance and should therefore, according to sound logic, have been subject to doubt. But "totality," "first causality," "origin"—concepts so charged with history, encumbered by so many theological associations, discredited by so many futile controversies—were regarded as "metaphysical." They were banished in principle from the thinking of the reasonable scientist as useless and dangerous. It should be added that investigation of the astronomical universe revealed such an immensity that to dare encompass it with the mind seemed to partake of speculative megalomania; besides, despite the warnings of Kant, the *infinity* of the universe was more and more accepted by science and professed as veritable dogma.

If there is, then—and we shall try to show that there is—a cosmology between the end of the seventeenth century and the end of the nineteenth century, it becomes ever less explicit and, as it were, apologetic. In any case, it is presented less as a science and more, in the nineteenth century, as a system of limiting assumptions; some accepted without criticism, others proposed or suggested at the frontier of knowledge and with all due reservations.

But once again the perspective changes totally at the beginning of the twentieth century under the impact of the greatest physicist of the age, and cosmology returns by entirely new paths to the domain of science, overthrowing the barrier raised against it in the classical age. We shall try to understand why and how modern science has once again received the universe into the company of acceptable concepts, without having lost any part of the strict discipline that protected science not only from the indulgences of the imagination but even from the excesses of rational speculation.

We have asked the question, Is cosmology a science? and we have answered it by a long history. The fact is that for such a question there is no other kind of answer; for the meaning and importance—variable, as we have seen and as we shall better understand—of the cosmological enterprise are, in fact, inseparable from the meaning and importance which have been assigned to the scientific enterprise, the enterprise of understanding nature and mastering it by means of reason. With regard to the value and significance that have been assigned to cosmology, it is not the content of scientific knowledge or the results of science which are in question, but science itself and the destiny of reason; in other words, we are dealing with a philosophical question.

PART ONE

THE COSMOLOGY OF THE ANCIENTS

Philosophical Account

Is it possible to recover the perfectly naive gaze that could have been directed at the world by a primitive human being? This appears, at first glance, quite improbable; modern sociology has made short work of the "natural man" whom the philosophers of the eighteenth century generously endowed with all the prejudices of their age. But let us look more closely. For if it is arbitrary to attribute to "natural man" ideas about life, the family, love, work, the gods or justice (all things which one must first have named in order to be able to think about), it is no less arbitrary to suppose that ever since prehistoric times men have been seeing, hearing, and touching things just as we do. Did the cave painters of Lascaux people the universe with gods or spirits? Did they imagine an abode for the souls of the dead? Did they regard incest as monstrous, commendable, or unimportant? Were they monarchists or democrats? These are so many idle questions. On the other hand, who would doubt that they saw deer the way we see them—that they retained in their memories, as we do, a recollection or image from which they were able to represent these animals.

Did they confine themselves to looking at animals? Did their curiosity extend to the horizon and to the heavens? We have no idea. What we can reasonably surmise, however, is that if these primitive men devoted the same attention to observing the heavens and the earth that they paid to watching deer in motion, what they saw there would surely differ little from what we can see, make note of, and remember—if we are careful to forget what we know about Latin, geometry, or astronomy; and also to forget—perhaps most difficult—that we have watches and that we live in a social world where everything is carefully timed.

It is then possible to see taking shape before our eyes a system of cosmic appearances that is stable, permanent, and, in a sense, definitive—for it is certain that the sky we see does not change any faster than the deer, and doubtful that it could be otherwise with the eyes and brains of human beings. It is this system of appearances that served as a point of departure for the enterprise, and led it, almost forcibly, to its first coherent realization.

Seeing a deer and carefully observing its movements is not enough to develop the concept of painting it, still less to be capable of carrying out this project. Neither is seeing and observing, even systematically, the appearance of the heavens and the earth enough to project a cosmology onto them. Something else is required which takes us quite out of the realm of the certain and permanent data easily accessible to the naive gaze. What is required is the *idea* of an objective order which these appearances manifest, without its necessarily being identified with them; and we must have the intention of stating this order, of expressing or representing it. This intention is, properly speaking, *rational*, even if it is first carried out in the form of a story or a myth—and it is very difficult to say when, how, and why it first began to appear. Even if we could recover the naive view of the world, we certainly could not recover as simply in its nascent form the intention to describe and understand the world.

At most we can, with the help of the incomplete and ambiguous evidence provided by history, watch the cosmological intention emerge, take form, and become clarified and refined through its early manifestations. The outstanding fact is that in a certain period (the fifth through the fourth century B.C.) in a certain region of the world (the area of Greek civilization) a certain geometric image of the world took hold. A system was established which for a very long time (until a certain threshold of precision had been crossed, until a certain requirement of experimental rigor had been introduced into the science of nature) achieved the most perfect possible harmony between the rational intention and what the attentive eye discloses of the world. The important thing is to understand simultaneously that this system—which lasted twenty centuries—is totally erroneous and that it nevertheless carries a kind of permanent truth, in the sense that its economy results from a very subtle adjustment between what the eye sees, what the imagination constructs, and what reason requires.

I. THE EARTH, THE SKY, THE HORIZON (Ia, b)*

The earth, the sky, the horizon: these are the first elements of cosmology at the level of perception. They are what is revealed once we succeed in disengaging ourselves from that chaotic flow, that stream of actions and reactions, of looks and gestures not completed, of germinal thoughts, of limited explorations, and constant readjustments in this familiar and ever-changing world that surrounds us, in which each thing beckons to us and slips away, but sends us back incessantly to the quest. Is this the *world?* Yes and no; the world is certainly here right beside us, all around us: this table, this house nestled at the crossroads, and the confused noise that rises from the street; but it is also further, higher, lower, beneath the floor, behind the house;

* See the explanation of the reference system on page x.

beyond this noise there are men talking, motors turning over, the wind blowing; beyond the town, the plains and beyond the plains, seas and mountains; the world is all this, but what about the world *itself?* Can one see it in itself? Can one, by rising above all these infinite details that keep surging up with each moment, with each movement of the hand, the eyes, or the imagination, can one contain it with a single look, surround it with a single gesture, embrace it with a single thought? This is a philosophical question which does not come naturally to the mind of someone who busies himself in the world and has to deal with things, animals, or people.

Let us assume, however, that there are situations which by their very nature give rise to the question and suggest the beginnings of an answer. If we are standing on a high place overlooking a flat landscape, and if we imagine ourselves to be wide awake and at leisure, in pursuit of neither a real rabbit nor a personal fantasy, accepting what offers itself to our view with a rather detached and nonexclusive interest, it is the world itself that will be presented to us in its primitive unified appearance, the inevitable point of all cosmological syntheses.

It does not matter whether it is day or night, provided the night is not too dark and neither fog nor snowflakes obstruct our view of what is before us. Then the infinite and infinitesimal details whose confusion and succession habitually form the texture of the world automatically reveal their insignificance; from the foreground into the distance, objects melt imperceptibly into one another. Almost all, all those that we touch, that we could touch, flatten and congeal into a dense continuous mass, the ground, the earth.

The earth is below, retaining against its surface everything that has some consistency; we too are below, and the earth hardly ever allows us to leave it for the sky, severely punishing our distractions, our imprudence, our fantasies in this regard. Even in dreams, in which we play delightful games with gravity, we do not really leave the earth, we simply take the liberty, like birds, of choosing the moment of our return, instead of being summoned back without warning.

Down below, beneath us, the earth extends, indefinitely perhaps; but in this direction, too, exploration is not very natural. Anyone who wishes to go down into the earth is always in the situation of finding himself stripped of his supports, always runs the risk of being swallowed up in darkness. In all cosmologies, the world under the earth remains opaque and somewhat mysterious; this is also true of our own, which probably knows more about what goes on in the sun than in the center of the earth. One of the great victories of rational thought may have been to circumscribe the mystery, at least, by conceiving of the earth as a sphere, an idea which, for the past twenty-four hundred years at least, educated people have not doubted. But this is not at all what we see if we return to first appearances.

What we see is a great flat disc on which things seem more and more squeezed together the further we look. The tree which is next to us shoots up toward heaven, defying the call of the earth; its counterpart a few hundred yards away is no more than a tuft of moss on an unbroken carpet; further still, the tree blends into the ground. Even the steepest mountains are flattened this way by distance; the horizontal gradually consumes the vertical, and at a distance everything becomes level.

Above us, on the other hand, the sky presents a very different aspect. Even this is an understatement: indeed, as natural observation is made precise and refined, the idea takes hold—false in fact, but so reasonable in appearance—that heaven and earth are contrary elements whose association forms the universe. Whether we look at it in the daytime or at night, the sky is like a dome. Toward the top, as toward the bottom, the world seems to close, but in another way. Between top and bottom yawns space, unfathomable, impassable; a gap ambiguous, surprising, disquieting, about which man apparently wondered well before and apart from any attempts at rational synthesis. According to a very popular saying, the Gauls had doubts about the solidity of the celestial vault; the Babylonians, on the contrary, believed that the gap between the sky and the earth resulted from their separation.

However this may be, if we continue to go by the first appearance of the world, the geometry of the sky is manifestly different from that of the earth. Out of this celestial vault of indeterminate shape, the astronomers made a sphere, thus expressing two results of observation that are undeniable and very important for cosmology:

▪ If our gaze, starting from the zenith (that is, from the point in the sky located directly over our heads), moves away from this point continuously, it descends toward the earth.

▪ Celestial objects are inaccessible, and located at indeterminate distances, which are probably very great. To locate them one can only measure *angles*, or intervals of time, never distance; it is as if they were all the same distance away.

Let us discuss these two points, but in reverse order, for we shall have to return to the second, i.e., what are we to understand by "celestial objects"? To the naive observer, clouds and lightning are in the sky, just as much as the stars are; but closer observation would soon show us that this meteorological layer* which lies between the earth and the "true" sky belongs more certainly to the kingdom of the earth than to that of the sky. It sometimes happens that the clouds touch the earth, lightning too, and the caprices of the meteorological world stand in contrast to the regularity of phenomena that are properly speaking celestial. If, moreover, we deliberately leave aside those questionable objects, the shooting stars, we shall then use the term "celestial objects" to refer to the *stars*, that is, the sun and all bodies that shine in the night.

Now, these celestial objects, properly speaking, are obviously separated from the earth by an interval which is physically impassable (aviation in no way changes this situation, and space travel, for the moment, changes it very little) and of indeterminate expanse. If, from our observatory, we locate a point on the earth's surface, nothing, *a priori*, stands in the way of our crossing the distance that separates us from it. Land or water will carry us to that point. Even if we do not physically cross the distance, our eyes can at least make use of landmarks (it is more difficult, of course, at sea) which will enable us to make a rough

* In Greek mythology the idea of the intermediary position of meteorological phenomena between heaven and earth is well represented by the personality of Zeus. This king of heaven frequently visited the earth, drawn by the charms of mortals; and he was the god of rain, clouds, and thunder.

estimate which is at least approximate and relatively determined (the woods are less far away than the village, the big oak is no more than 200 feet away, the boat is passing behind the island, etc.). On the contrary, between the true sky beyond the clouds and ourselves, no contact is established or seems capable of being established. Everything is uniformly inaccessible, and, even in a purely qualitative way, it is very difficult, without minute, patient, methodical observation, to locate objects at different relative distances. To assure oneself by observation that the stars are further away than the moon is already a problem for scientific discovery.

THE HORIZON

The most important result of ordinary observation, the one first cited, is the meeting between the celestial vault and the terrestrial disc along that circle, so visible and impressive, which marks the common boundary of heaven and earth, the *horizon.**

For one who seeks to understand appearances, who hopes thereby to reconstruct a rationally satisfying image of the world—for one, in short, who would be a cosmologist—the comparison of these two phenomena immediately gives rise to a mystery. If it is true that the sky and the earth meet at the horizon, how can they retain their different natures, how can they maintain their impassable distance? How can the sun, which is swallowed up every evening by the earth or in the sea, reappear intact, in all its glory, eternally faithful to its daily round? Here again, mythology indicates that the problem was recognized and that solutions were proposed well before the appearance of a rational cosmology. Thus, in the Homeric period (around the ninth century B.C.) the Greeks imagined the terrestrial disc as surrounded by the river Ocean, which marked a transition between heaven and earth.

The "true" solution to the problem of the horizon everyone now knows, of course, for tradition excuses us from having to discover it for ourselves: the horizon is only an appearance which

* The word "horizon" comes directly from the Greek. It is a transliteration of a verb that means to bound or to limit.

recedes forever as we try to approach it. But to understand this is to understand that the earth is not a disc or a cylinder, but a kind of ball. This means that of the two contradictory appearances of direct observation, the spherical sky and the flat earth, the first is more true than the second. In terms of the practical certainties of daily life the reversal is of importance, for how are we to believe that it is this earth, hard and demanding, but reliable, that is lying, and the sky, distant and inaccessible, that is telling the truth?

THE TERRESTRIAL BALL AND THE SPHERICAL UNIVERSE

The discovery of the curvature of the earth* must be regarded as an event of primary importance in the progress of cosmology, for it established definitively the dominion of the sky in the investigation of the universe, and it marks the first step toward the establishment of a *spherical* concept of the universe. We shall try to understand how this concept, to which the whole first part of this book is devoted, came to be constructed and why it flourished for so long, despite its falsity.

Where, when, and how did man first understand that the earth is an enormous ball? We wish we knew. A horizon that recedes is not a thought that would occur to a sedentary peasant, for whom the fixed, familiar, and sure landmarks of the horizon announce rain or drought, autumn or spring. The nomad, the caravaneer, and the navigator, on the other hand, are better prepared to notice that the earth and the sea disappear and reappear in a new form indefinitely, and also to rely on the sky when the monotony of the landscape robs them of all landmarks. It is not

* It is, of course, quite a distance from the vague hypothesis that the earth is a ball to the precise, though strictly speaking, false hypothesis that it is a perfect *sphere*. But for the Greeks, who were first of all geometers, the geometrical form of the sphere, because of its exceptional properties, naturally presented itself to clarify the vague idea of the ball. Among the numerous proofs that astronomers of antiquity have given for the roundness of the earth, few in fact prove that it is a *sphere*. According to the distinguished historian and philosopher of the sciences P. Duhem, the best is one of those that Aristotle gives: the shadow of the earth on the disc side of the moon, at the beginning and end of eclipses, is always circular in shape.

surprising that the Near East with its deserts and Greece with its islands provided the West with its first astronomers and cosmologists. In any case, an ancient tradition attributes to the celebrated Pythagoras (the perhaps legendary founder of a school that flourished in Italy in the late sixth and fifth centuries B.C.) the first theory of the universe as *spherical* and *geocentric,* according to which *the universe is a sphere in the center of which is fixed, immobile, the terrestrial sphere.*

Observation of the interior of the earth, like observation of the ocean floor, has remained extremely limited up to the present. The fragmentary results provided by the exploration of caverns, the digging of mines, the diving of fishermen, and the observation of volcanoes have certainly done much to inspire the imagination of men; but these findings, precisely because of their fragmentary and uncertain nature, do not lend themselves to the kind of cosmological synthesis which, since preclassical antiquity, has been naturally associated with the observation of the heavens. When, at the beginning of the fourth century B.C., Plato wrote his *Timaeus,* the first systematic treatise on cosmology that has survived in its entirety, astronomy and cosmology had already for thousands of years, perhaps since the beginning, been a single science. And so it is now toward the heavens that we must, from our improvised observatory, direct our attention.

II. THE STARS

Only the stars concern us here, and among them, only those which are visible to the naked eye and which, in addition, appear with sufficient permanence or regularity to belong to the physiognomy of the universe. This results in a rather heterogeneous family, whose members we may as well list at once: the sun, the moon, the five planets (Mercury, Venus, Mars, Jupiter, Saturn), the stars (about six thousand).* One can group these stars in various ways, and each of these groupings has a certain importance from the point of view of cosmology. One can group them:

▪ *According to brilliance,* in which case the sun, a diurnal

* We refer, of course, to those planets and stars visible to the naked eye.

star, is distinguished from all the others, which are nocturnal stars.

▪ *According to appearance.* The sun and the moon are the only ones that appear as discs. The others, planets and stars, appear as luminous points.

▪ *According to their apparent motion.* Another system of classification which is by far the most important from the point of view of cosmology. In contrast to the "fixed" stars, which move *en masse*, there are the seven "wandering" stars: the sun, the moon, and the planets; it is to this characteristic that the planets owe their name.*

(III) All the stars "shine." Although, in fact, all are not, each seems to be the source of the light that emanates from it. But the brilliance of the sun so exceeds that of all the other stars that the difference between day and night, to which the behavior of a great many living creatures is attuned, is intuitively understood as a qualitative difference, and even an actual opposition; very early the tendency of our thinking toward dualistic patterns seized upon this opposition. Mythologies, religions, and philosophies of the most diverse kind bear witness to the importance assigned by imagination and reason to the opposition of day and night, with which have been associated in a thousand ways the other most striking pairs of opposites: male and female, good and evil, life and death. . . . Although the difference between day and night has regained, in the discussion of certain modern theories, an importance which it had lost in the classical age, it is still no more than a great quantitative difference between a weak light and a strong one; dawn, twilight, moonlight mark the transition. Even Venus sometimes wears a shadow; whoever first clearly understood that the difference between day and night was cosmologically insignificant was certainly a penetrating mind; we do not know who he was. Plato, at any rate, knew where to start. He explicitly puts day and night in their places within the order of appearances: in the making of the world by the "designer," "God," or "demiurge" (terms he used interchangeably),

* "Planet" comes from the Greek verb meaning "to wander."

the creation of the sun, and hence of day and night, marks the end of the work that can properly be called cosmic. It has primarily an educational value, for it enables living creatures to form an idea, by means of sight, of the mathematical law that governs celestial movement.*

(Id) If we turn our attention to the appearance of the stars, we have on the one hand the sun and the moon, and on the other hand the group consisting of planets and fixed stars, which are like luminous points. We know now that this circumstance does not signify an important characteristic of the astronomic world. The pinpoint appearance presented by most of the stars is due to their very great distance in relation to the capacities of the eye, whose biological functions are performed at much shorter distances. The important consequence of this state of affairs is that the eye somehow pushes back into infinity almost all the stars without distinction, whereas their actual distances are extremely varied. We see both Sirius and Venus as very brilliant points of light; but we know now that the first is millions of times more remote than the second.

But we must forget this in order to understand how the naive observer was able to construct an image of the world. For him—for us, in our improvised observatory—the stellar world, the world of luminous points, has qualities that are truly extraordinary in comparison with everything that we can see on earth, qualities that confirm the idea that the sky is decidedly something else. Pure light, without extension, without volume, without consistency: these stars are in some sense at the limit of that which is material, at the frontier between the world of signs and the world of things. To be sure, they are subject to space, to number, to time; but this is all they have in common with ordinary physical objects. They provide our senses—only one of our senses—with only what is needed—or barely a little more—to calculate their positions, to count them, to classify them. To this is added the fact that, as we shall see, the permanence of their relative positions (their "fixity" which in actuality belongs strictly only to

* *Timaeus,* 39*b.*

the fixed stars), the constancy of their light, the regularity of their motions—all these qualities seem to place them beyond the contingencies and disorders of earth. For minds given to seeking the order of reasons and numbers, the stellar world is like a vast luminous mathematics to be deciphered; it is also like the promise that the effort to understand things is not a hopeless one. "Astronomy," wrote Henri Poincaré, "has given us a soul capable of understanding nature."* It is not surprising that the word "divine" was commonly used in classical antiquity to describe the stars. There are four species of living things, according to Plato; the first is the celestial species of the gods, whose form is composed of fire.† It was a long time before the gods deserted the sky of stars, and this empty sky, by his own admission, frightened Pascal; it is obvious that gods visible in the sky are more reassuring than the god hidden in the depths of the soul.

CHARACTERISTICS OF THE APPEARANCE OF THE STARS (Ic; III)

Let us return to the appearance of the stellar world and its remarkable features. Composed of single dots of light distributed over a sphere, it spontaneously invites mathematical description because it can be reduced to a limited number of numerically expressible *parameters:* position, motion, brilliance.

With regard to *brilliance*, the attentive observer can easily distinguish one star from another; less easily with regard to color. In any case, for each star these two parameters are, on the human scale, practically constant within the domain of stars visible to the naked eye. Of course there are exceptions in the case of brilliance, that is, variations, of diverse amplitudes and durations, which we know to attribute to causes which are also diverse (greater or lesser distance of the planets, mutual occulation of two stars in the case of double stars, intrinsic variations in luminosity, which can be quite spectacular in the rare cases of novae and supernovae); but for the observer without instruments, the effect of *permanence* is overwhelmingly dominant. For example, in the winter sky the brilliant stars Sirius and Betelgeuse

* *La Valeur de la science,* Chapter 7, page 157.
† *Timaeus,* 40.

immediately provide a reliable and familiar landmark. If each star maintains its brilliance it is possible by careful comparison to classify the stars in order of decreasing luminosity. Astronomers use the term *magnitude,* and calculate in the other direction: i.e., magnitude increases as brilliance decreases; a very brilliant star is of the *first magnitude;* the eye sees down to the sixth magnitude. The value of this classification, which was originally based on purely visual observation, rests on an implied hypothesis—namely, that the properties of the human eye do not vary greatly from one individual to the next, and that they are governed by rather simple laws. The eye is a rather good instrument of detection for luminosity; modern classification has not overthrown the traditional classification, but has refined it considerably while still preserving its principle.*

If the naked eye can reasonably classify the stars according to brilliance, it cannot, however, classify them numerically according to color. At most it can note important qualitative differences. For example, the difference between the planet Mars, red, and the star Rigel, blue, is immediately noticeable; it requires a certain degree of attention to notice that Arcturus and Vega, two brilliant stars rather close to one another in the summer sky, have distinctly different colors. Numerical classification of the stars by color requires use of a spectograph.

The most striking constancy in the apparent world of the stars is the constancy of their relative positions on the sphere, their grouping into stable configurations, the *constellations.* Some of these, like the Great Bear, Orion, and Cygnus, which combine several brilliant stars in very characteristic designs, retain a strange power of fascination. The stability of the constellations is, even more than the constancy of stellar luminosity, an appearance so convincing and presupposes such a high degree of precision in observation before one could call it seriously into doubt, that it has remained to the very end the best argument of those who believe in a spherical universe. It was only after the collapse of

* This principle represents the spontaneous application by observers of the psychophysical law of Weber-Fechner, which so many philosophers have discussed without understanding its meaning: the relationship between actual intensity of the radiation received and apparent brilliance is a logarithmic one.

geocentric cosmology that astronomers were able to prove that in actuality the constellations must slowly be losing their shape. By way of contrast, the discovery of variations in brilliance in the stars helped to destroy the confidence of the great astronomers of the sixteenth century (who had the good fortune to be able to observe some very brilliant novae) in the traditional image of the immutable heavens.

The planets, on the other hand, together with the sun and the moon, form, by virtue of their apparent motions, a group that is quite distinct from the fixed stars. This category, unlike the preceding ones (day star and night stars, pinpoint stars and disc-shaped stars), is of decisive importance from the cosmological point of view. As astronomers finally came to understand, these stars all belong to a single system of bodies of which the earth is a member and which are related to the sun by universal attraction. They are all, relative to the fixed stars, very close to the earth; a fact which does not prevent the planets from resembling the fixed stars in their pinpoint appearance. On the other hand, astronomers noticed very early that the sun, the moon, and the planets are distinguished as a group from the fixed stars by the characteristics of their apparent motion in the sky.

THE SPHERE OF THE FIXED STARS AND DIURNAL MOTION (Ib; II)

The fixed stars revolve, but as a unit, without the constellations losing shape, as if they were fastened to a solid sphere turning around an invariable axis. In the northern hemisphere, where all the astronomical observations known up to the Renaissance were made, this axis is easy to locate. In fact, its direction coincides almost exactly with that of a rather brilliant star, Polaris, which alone in all the sky seems completely immobile. The sphere of the fixed stars makes one revolution in a day; precisely because of the special qualities of the sun's motion, this *sidereal day* gradually shifts in relation to the alternation of day and night, which depends on the solar motion.*

* In comparison with the twenty-four-hour day of our calendar—an artificial unit of time based on the sun's motion—the motion of the heavens advances by about four minutes a day.

The uniformity and regularity of the motion of the heavens is of great importance from the point of view of cosmological thought. It provides reason with a model of regularity and periodicity, a model of uniform time, in short, with an intuition of an intimate kinship between time and number. It must have been after meditating on the periodical rotation of the heavens that Plato arrived at his famous definition of time as "the moving image of eternity."* Together with the remarkable properties of the fixed stars—which caused astronomy among the Greeks to be a science dominated by geometry—the apparently perfect uniformity and periodicity of celestial rotation dictated the idea that numbers dominate the science of the universe, an idea which, as we know, has had a long success.

MOTIONS OF WANDERING STARS (Ic; II)

But although the sun, the moon, and the planets, that is, the most brilliant stars in the sky, revolve along with the sphere, drawn along in its diurnal motion, they are not fixed within it; they wander from constellation to constellation. This wandering is itself orderly, and in two senses: (1) The seven wandering stars all remain confined within a certain limited area of the sphere. (2) Their motions, although much more complex than those of the sphere of the fixed stars, are nevertheless themselves periodical.

1. This special area, which extends on either side of a great circle of the sphere and in which all the wandering stars are always found, has been known for a very long time. Called the zodiac, it contains a group of twelve constellations of which some, like Scorpio or Leo, are easily recognizable, and others, like Cancer or Libra, are much less so. These constellations have traditionally been named and symbolized by signs which are encountered in all representations of the zodiac in very different eras and in very diverse regions. Sagittarius or Gemini can be recognized at a glance just as easily in the representations of India

* *Timaeus,* 37*d.*

as in those of medieval Europe. The language of astronomy is international, and may have been for thousands of years, but little is known about its origin.

2. Despite the first appearance of great complexity, gradual and laborious discovery uncovered a whole network of periodicities behind the confusion of planetary motions and led cosmologists, throughout the first phase of the theory of the universe, to believe that in the celestial world all motions are periodical, and that uniform rotation is the fundamental periodic motion. Indeed, this postulate came to be accepted as an absolute principle of the science of the universe. The reason is that this concept, in its rational simplicity, has proved to be extraordinarily effective, in spite of the increasing difficulties that have been encountered in applying it, in making order out of the more or less complicated appearances presented by the motions of the planets.

The regularity of the moon's motion—so striking because it is accompanied first by the gradual unveiling, then by the occultation of that luminous body in a rather short interval*—is easily observable and has certainly been known everywhere from the earliest times.

The same is true of the regularity of the sun's motion, which has much more immediate consequences for practical life, since it is this cycle that governs the alternation of the seasons. But the first appearance of this motion is even more complicated than that of the moon's motion, and provides a first example of the effectiveness of the postulate of uniform rotation. In relation to the sphere of the fixed stars, which naturally draws it along in its rotation, the sun shows a double motion: it falls back, in relation to the fixed stars, by one complete revolution in a year (this is the definition of a year); and it rises for six months and falls for six months, which is to say that the elevation of the highest point it reaches above the horizon swings back and forth between

* In relation to the stars, the motion of the moon is very noticeable, as is the narrowing or widening of its crescent.

the two *solstices;** at the same time, the point where it rises and sets shifts along the horizon.

But this double motion can be reduced to a single one if we agree that in relation to the sphere of the fixed stars, the sun travels in one year a great circle, the *ecliptic*, whose plane intersects that of the equator at a fixed and determined angle. This revolution is not, in point of fact, uniform, but the deviation with respect to uniformity is not easily detected. Interpretation of the sun's motion by its revolution on the ecliptic is, as noted, the first example of applying the fundamental method of interpreting celestial motion in geocentric cosmology, the method that follows naturally from the postulate of periodicity and uniform rotation. The method consists in interpreting the motions of the planets, however complex they may be, as a combination of circular motions. The essentially temporal notion of periodicity has been closely associated with the notion of circularity, as is rather well indicated by the word *period*, which means in Greek "way round" or "circumference."

The temporal adjustment of the three fundamental cycles of the heavens, *day* (period of the fixed stars), *month* (period of the moon), and *year* (period of the sun), soon raised for astronomy a difficult problem to which we now know that there can be no exact solution. Actually, the three periodical physical phenomena which are behind these appearances (rotation of the earth on its axis, revolution of the moon around the earth, revolution of the earth around the sun) are certainly not independent, for they are probably related by origin (they also follow the same direction). However, they are not related in the sense that there is any simple numerical relationship between their periods. The adjustment between day, month, and year can only be made in an approximate way, as the peculiarities of our calendar inform us (months of 28, 29, 30, or 31 days, years of 365 or 366 days). And yet it was rather natural to believe in a harmony among the celestial periods, to assign to the regular motions of the planets a greater simplicity than they possess. Especially since, given its

* According to etymology, the solstices are those moments in the year when the sun *stops* rising or falling.

effects on the measurement of time, the adjustment of cosmic periods has social and political consequences. In Egypt, from the time of the Old Kingdom, the calendar was a matter of state. It is the name of Julius Caesar—and not the name of the astronomer Sosigenes, who recommended it—that remains associated with the calendar that provides the backdrop for the whole history of Europe until the end of the sixteenth century; and the revised calendar by which we live bears the name of a pope, Gregory XIII, and not the name of a cosmologist.

THE MYSTERY OF THE PLANETS (Ic, d; II)

But the real difficulties, for an adult astronomy and for a deliberately rational cosmology, begin with the motions of the planets. These stars (as far back as one goes, astronomy knew the five that are visible to the naked eye*) are very easy to observe (except Mercury), and even brilliant (notably Venus and Jupiter), and yet at the same time very confusing. They move along the ecliptic in a very capricious manner, and there are striking differences in motions from one planet to the next. While Venus, like Mercury, seems to be associated with the sun, which it follows or precedes at most only by about four hours, the others may occupy any positions at all along the ecliptic in relation to the sun. All of them rotate in relation to the sphere of the fixed stars according to a motion that seems very irregular, advancing more or less quickly, sometimes stopping and then starting off again in the opposite direction. Their relative positions are seemingly unlimited; they may at times pass behind the moon, which proves at least that they are further away. So strong, however, was the belief—at least among the Greeks—in the necessary periodicity of events in the heavens that the theory of the "Great Year" (the hypothesis that there exists an interval of time at the end of which the universe finds itself in the same state as it was in the beginning) occurred to them rather early, perhaps because of the contrast between the simplicity of

* Mercury, Venus, Mars, Jupiter, and Saturn. Classical and modern astronomy have added to the list, in addition to the earth itself, Uranus, Neptune, and Pluto, all three beyond Saturn, and numerous small planets which gravitate between Mars and Jupiter.

the motion of the fixed stars and the complexity of the motions of the planets. Plato, at any rate, gives a strictly astronomical definition of the "Great Year" by associating it directly with the motion of the planets which must periodically find themselves, he thought, in the same position.* Of course, he was far from having the means to prove this regularity (which even the theory of gravitation cannot prove with exactness) and *a fortiori*, to calculate the period.

III. THE SPHERICAL WORLD AND THE REASONS FOR ITS SUCCESS

We have listed, in a somewhat informal manner, all the fundamental astronomical facts that might appear to an observer provided with nothing but time, patience, and good eyesight. Egyptian, Babylonian, Greek, and Arabian astronomers had more than this, notably instruments with which to measure angles. On the one hand, they had no telescopes (that is, no instruments that enlarged the apparent diameter of distant objects); on the other hand, all their measurements were superior to the general level of precision that was so crude in comparison with what modern astronomy can reach. Consequently, in spite of the demonstration of certain very subtle phenomena—like the precession of the equinoxes, discovered by Hipparchus (Ic2)—or the execution of difficult measurements at a respectable order of magnitude—like the measurement of the radius of the earth by Eratosthenes or the measurement of the distance from the earth to the moon by Aristarchus—scientific astronomy did not succeed in essentially modifying the physiognomy of the heavens as it presents itself to the unaided observer until Galileo. This is a circumstance that must be kept constantly in mind in the discussions that follow.

Let us now review the elements we have so far been able to assemble for our system of the world and the hypotheses to which our observations have naturally led us:

The sky is spherical, and the stars with which it is filled are inaccessible and certainly very remote. The earth has a regular

* *Timaeus*, 39*d;* according to Duhem, the theory must have been borrowed by Plato from Pythagorean teachings, notably from Archytas of Tarentum.

shape, in spite of its apparent irregularities, and the receding of the horizon, together with other signs, proves that it is not flat, that it is a sphere in the center of one's world, concentric with the sky. Gravity keeps it compressed together and creates a fundamental opposition along all verticals between top and bottom. But these verticals are not parallel; they radiate outward from the center of the earth toward the sky.

Celestial objects are qualitatively different from terrestrial objects—simpler, purer, apparently immutable in their incessant revolutions. The fundamental motion of the sky is in perfect agreement with its spherical form: the sphere of the fixed stars rotates as a whole around an axis which joins the earth to a fixed point in the sky. This regular and uniform rotation, which gives a constant rhythm to the course of existence, is what we call time.

However, certain heavenly bodies, and not the least significant, have more complicated movements. Although drawn along by the movement of the sphere, they are nevertheless not fixed within it. To this category belong the moon, the sun, and the planets, whose capricious motions are a mystery. Moreover, since eclipses show that the moon passes in front of the sun, and that it is possible to see the planets disappear behind the moon, it is difficult to conceive of the heavens as a *single* sphere.

It is here that the division arises between the first and the second cosmology. Henceforth one has two alternatives:

▪ One may argue, first that the spherical model so clearly suggested by observation is still the right one, by assuming that the heavens consist not of one sphere, but of several, each one centered more or less exactly on the earth, and each one moving in relation to the others. It was this solution that the majority of philosophers and astronomers adopted until the sixteenth century.

▪ Alternatively, one may assume that, contrary to appearances, the world is not spherical, that the earth is not in its center, that the earth is not immobile, but that it revolves upon itself and around the sun. This is the modern solution, that of the second (and third) cosmology.

For us, who have behind us Copernicus, Galileo, Descartes, Newton, and three centuries of rigorous astronomical measurement, it is no longer possible to doubt that the first cosmology is incorrect; and we are amazed that thinkers of the stature of Aristotle or astronomers of the caliber of Hipparchus and Ptolemy could have been taken in by it. Indeed, this apparent credulity deserves examination, and a little history will reveal the surprising circumstance that in fact the Greek astronomers understood very early that geocentrism was not an indisputable system of the world, but that nevertheless geocentrism was generally accepted. We shall have occasion to ask ourselves why so many great minds never succeeded in understanding this thing that is so simple to us, that the celestial vault is only an illusion that recedes indefinitely above an astronaut who rises in the sky just as the horizon recedes before a navigator.

We recall that tradition attributes to Pythagoras the first explicit formulation of geocentric cosmology. But it was a disciple of Pythagoras, Philolaos, who, according to this same tradition, is said to have been the first to doubt the fixity of the earth and its central position. Philolaos thought the earth revolved around a central fire which men, because of their position on the earth, had not been able to observe. Imagination obviously played a greater role than science in the theory of Philolaos; this cannot be said of the theory constructed two centuries later by Aristarchus of Samos. Less than a century after the death of Aristotle (that is, 1,800 years before Copernicus) this astronomer correctly formulated the hypothesis that the earth is a planet like the others, rotating on its own axis and revolving around the sun at the same time. This hypothesis correctly explains both the rotation of the fixed stars, an appearance caused by the rotation of the earth upon itself, and the complexity of the motions of the planets, an appearance caused by a multiplicity of actual motions.

The work of Aristarchus by no means went unnoticed by his contemporaries and successors. The famous mathematician and physicist Archimedes, in particular, understood perfectly that given the absence of parallax* of the fixed stars, the hypothesis of

* At intervals of six months we observe the stars from two different points that are separated by a distance equal to the diameter of the earth's orbit. In relation to a fixed direction, the angle from which we see a star must

Aristarchus implied that the stars were separated from the earth by enormous distances, disproportionate to the distance from the earth to the sun.* Unfortunately, Archimedes did not place his genius at the service of cosmology. The problem interested him only in its mathematical form, i.e., if the stars are as remote as Aristarchus' hypothesis implies, how are we to conveniently express the *numbers* that measure these distances in terrestrial units? The modern reader is a little disappointed to find a genius of this magnitude (who in certain respects was several centuries ahead of his contemporaries and whom Galileo would one day use as a model of reasoning in physics) unconcerned with the fundamental problem, namely, was Aristarchus right or not?

But it is a fact that not one of the great scientists of antiquity or the Middle Ages retained the hypothesis of Aristarchus. All were satisfied with the geocentric theory, to which Ptolemy gave its most perfect form in the second century after Christ. Why this long fidelity to a model of the universe which we are inclined to regard as "naive," when the other solution, the solution of the future, had been explicitly offered to them—when, moreover, the "orthodox" geocentric model presented difficulties of interpretation that increased in proportion to improvements in the precision of measurement?

To a question of this kind there probably is no sure answer. But at least one can begin by dismissing certain historical illusions caused by the particular circumstances surrounding the decline of the geocentric cosmology. Among these circumstances one of the best known is that the Catholic Church threw the weight of its authority on the side of the ancient cosmology and made it more dominant than ever at a time when all informed minds were convinced that it was false. Thus we are inclined to believe that the long persistence of this error was caused by the intervention

---

have changed. This variation is the *parallax* of the star. But the stars, as Archimedes realized, are so remote that few have a readily observable parallax and it is always very small; the ancients were unable to observe any at all.

* The profound accuracy of Archimedes' views on this point may be seen if one considers that the ratio between the diameter of the earth's orbit and the distance from the earth to the nearest star is on the order of 1 to 1,000,000.

in its favor of an intellectual and moral authority that put forth extrascientific, nonrational arguments. But this is a retrospective illusion. At the time that Aristarchus proposed the Copernican hypothesis in opposition to the cosmology of the spheres to which Aristotle had given quasi-definitive form a century before, the Catholic Church did not exist, and no institution exerted a comparable dominion over men's minds. Every man, provided he was educated, was free to say what he thought about Aristotle, just as Aristotle had been free to oppose his teacher Plato, especially in the area of cosmology. It was not the weight of tradition or the threat of persecution that prevented Aristarchus' contemporaries and successors from supporting his hypothesis and from assuring his success almost two thousand years ahead of the fact. If, indeed, the cosmology of the spheres was given the advantage of a nonscientific and extrarational support, that happened much later.*

The first scholars of Christianity regarded all of ancient cosmology with mistrust because they saw it as guilty of deifying the world and the stars, and they hardly bothered to defend one hypothesis against another. But toward the end of the Middle Ages, beginning in the fourteenth century within the universities (although they were dominated by the Church), numerous doubts began to appear and later were freely expressed concerning this or that aspect of Aristotelian philosophy. In particular, the thesis that the universe is finite was felt by certain theologians to be incompatible with the dogma of divine omnipotence.

Hence ecclesiastical orthodoxy did not assure the success of the spherical cosmology in the beginning, and only episodically and imperfectly protected it from decadence at the end. The

* Duhem has tried to accredit the opinion that Aristarchus was threatened with persecution because of his astronomical theories; but this opinion rests only on an accusation of blasphemy said to have been made against Aristarchus by the Stoic philosopher Cleanthus, who was known for his narrowmindedness; and Cleanthus' remark is known only through the testimony of Plutarch, who was writing several hundred years after him. The successors of Alexander, although they put an end to the political freedoms of the Greek city-states, did not impose an ideological dictatorship on their subjects. On the contrary, some, like the Lagides in Egypt, generously promoted the sciences.

lasting success of the first cosmology must therefore be attributed essentially to other, more intrinsic reasons, to the specific virtues that the spherical model of the universe possessed for minds which were both attentive to the findings of observation and convinced of the rationality of the world, in short, for *scientific* minds.

Let us then examine the very strong rational factors which—for an observer equipped only with rather crude instruments for the measurement of time and angles, and not possessing a telescope—militate almost irresistibly in favor of a spherical universe centered around an earth which is itself spherical and immobile.

At what point the distinction between the earth and the sky and the apparent sphericity of the sky become obvious to perception depends upon the structure of perception, as we have already observed. On the other hand, the discovery that the earth is round, which was probably contemporaneous with the very formation of western rationalism, marks the outcome of a first conflict that has been resolved. It is a conflict between perceptive generalization and geometric generalization; and it has been resolved in favor of geometry, by going beyond tangible appearances to a rational form which integrates them by eliminating them. For the observer sees the earth as a disc and must think of it as a sphere and must understand that, on his scale, the curve of the sphere is flattened; he must think of the earth according to the model of the sky.

As for the problem of *celestial motions*, the problem of interpreting these within the context of the spherical cosmology (a problem which in the long run would turn out to be insoluble), it must be understood that during the period when this cosmology was being established the mere possibility of a simple geometric interpretation of the most fundamental of these motions—*a fortiori*, the gradual extension of this interpretation to more and more complex movements—this alone was regarded as extremely significant. Did it not offer a shining proof that rationality is present within appearances and that the mind has the power to discover it with the help of simple geometrical constructs?

The principle governing the interpretation of celestial motions in the geocentric system is, in fact, directly suggested by the fundamental appearance of the heavens in motion, the rotation of the sphere of the fixed stars around the axis of the world. Later, astronomers tried to reduce all celestial motions to a system of circular motions and to demonstrate their cyclical nature. For the naive observer, neither the opportunity for nor the possibility of this reduction is *a priori* evident. It was certainly a great victory for geometric thought to reduce, for example, the capricious and complex motion of the sun to two circular motions (a daily motion and a yearly motion) whose cycles have a relationship which, if not simple, is at least clearly defined—two circular motions which, up to a very high degree of precision, have every appearance of uniformity. And to determine on this basis that the moon and the planets deviate little or not at all from the *ecliptic* (the great circle described by their annual revolution of the sun) was an obvious encouragement to continue exploiting the same hypothesis.

Plato expresses in his own way the idea that the reduction of celestial motions to a uniform circular form is evidence of a correspondence between the perceptible appearance of the heavens and the intelligible structure of ideas. The equator is the circle of the self, and the ecliptic is the circle of the other, which means that these two cosmic circles correspond to two fundamental elements of his dialectic* (i.e., the structure of logical thought by means of which existence can be understood).

It is not surprising, then, that it was within the intellectual environment of the Pythagorean and Platonic schools that the future of cosmology was determined. In order to continue to apply successfully the principle of circular and uniform rotation, two great difficulties had to be overcome. Cosmologists had to explain the extreme irregularity of planetary motions, and they had to account for the unavoidable distribution in the depth of the stars, a distribution which the results of observation undeniably support and which rules out the description of the sky as a single sphere. From the point of view of astronomy, the final stage in the enterprise of geocentric reduction was reached in the second century by an astronomer from Alexandria, Ptolemy.

* *Timaeus*, 38*c*–39*d*.

The "mathematical system" of Ptolemy—the *Almagest*, as the Arabs later called it—accounted for all the appearances known in his time by means of an extremely subtle and complex architecture of circular and uniform motion. It remained the last word on astronomy until Copernicus, who lived fourteen centuries later. Its extreme subtlety makes it appear fragile; but its geometric complexity crowns an edifice that rests on a solid foundation. This is why in order to destroy the cosmology of the spheres it was not enough to ruin the Ptolemaic superstructure, as Copernicus did; it was necessary, as Galileo did, to attack the very foundations.

The real masters of the spherical cosmology lived well before Ptolemy; they were the astronomer Eudoxus of Cnidus and Aristotle. Eudoxus was twenty years younger than Plato and twenty-five years older than Aristotle; according to tradition, he studied first with the Pythagoreans, then with Plato, and then in Egypt. Eudoxus reconciles the multiplicity of celestial motions with the principle of the sphericity of the heavens by imagining the universe as a system of concentric spheres, contained one within the other like a series of boxes, with the terrestrial sphere in the center. Each sphere is drawn along by the motion of the one that surrounds it, but it has in relation to the surrounding sphere its own motion of uniform rotation. The fixed stars are attached to the outermost sphere, which has only one motion and carries all the other spheres with it. Each of the wandering stars is attached to its own particular sphere; the spheres of the sun and the planets revolve around the same axis, which is perpendicular to the plane of the ecliptic. The sphere of the moon, which is closest to the earth, revolves around a special axis—for Eudoxus and his contemporaries knew that the circle of the moon's apparent motion is slightly tilted from the ecliptic; otherwise, there would be an eclipse at every full moon (Id).

Eudoxus had constructed a celestial mechanics very different from that of the moderns, although it too was very far removed from appearances. But on the one hand, the form, or the structure of appearances, remained intact. By postulating invisible spheres, Eudoxus was, of course, adding invisible elements to the system of the world, but he maintained the spherical scheme that corresponds to the immediate appearance of the heavens. And on the other hand, by extending the use of the spherical scheme, he brought about the triumph of a geometric form which our reason finds much more satisfying than the disorder of appearances on earth.

It is the system of Eudoxus that forms the astronomical base of the cosmology of Aristotle, which represents a synthesis of this astronomy and a physics which is compatible with it and is, for this very reason, incompatible with our own. This brings us to the second important reason for the lasting success of the geocentric cosmology. The first physics, that is, the first coherent rational theory of natural phenomena that was constructed, was almost exactly in agreement with it and in disagreement with the other hypotheses, notably that of Aristarchus.

When confronted by the ingenious mechanical construction of Eudoxus (or the even more subtle one of Ptolemy), the modern mind cannot resist asking, what are the *forces* that move these spheres and why do they behave in this way? But this question did not arise, or at least not in this form, for the contemporaries of Eudoxus and Aristotle, because they had a conception of causality that was different from our own, as we see clearly in the works of Aristotle, the first philosopher to submit the notion of causality to systematic analysis. For Aristotle the system of motions in nature is not without cause, but it depends in the last analysis on a "first mover," itself motionless, in which everlastingness, power, intelligence, and harmony are reconciled. The influence of this "first mover" is exerted directly on the heavens, to which it imparts the simplest motion that can be conceived, uniform rotation, which, like itself, is eternal. Indirectly, through the intermediary of the celestial motions (the sun's motion maintains life on earth), the influence of the first mover is also exerted on the earth; but less rigorously. For on earth phenomena are less regular, less well ordered than in the heavens; they depend partly on chance, which is to say that they partly elude reason.

Must this "first mover" be called "god"? Yes, but it is a being very different from the Christian God. In the first place, it is not the only god. Next, it presides over the order of the world, but it is not concerned with souls, and if it has an effect on them, it is as the model of an eternally harmonious existence that is offered for their admiration. On this point too, however questionable such a doctrine may seem to us, it should be noted that the cosmology of the spheres had the advantage of responding to certain expectations of common sense. For the observer of the sky, exposed, as men have always been (although in very different degrees) to the hazards of illness, famine, war, slavery, living their daily lives in the presence of death and decay, the spectacle of the celestial motions, indifferent to the "sound and fury" of earth, eternally the same, easily gives rise to the idea of an existence protected from the certainty of decline: "There, all is but order and beauty. . . ."

When confronted by the celestial motions, must we go beyond an analysis that reduces complex appearances to simpler and more harmonious movements? The founders of the cosmology of the spheres did not think so. The belief that the uniform circular motions are sufficient unto themselves, that they carry their own justification, and that any subsequent research that attempted to explain them would be useless—this belief dominated the first cosmology. It continued to haunt the thinking of all the great cosmologists of the sixteenth century; it was this persistence that made the revolutionary work of Galileo so laborious. In order to pass from one system of the world to another, it was also necessary to pass from one philosophy to another.

In the philosophy of causality that was established with the Galilean reform of physics, a single system of mathematical laws was to govern celestial phenomena and terrestrial phenomena. But in the Aristotelian philosophy of causality that was in agreement with the cosmology of the spheres, the physics of the heavens remained distinct from the physics of the earth. And if one is trying to understand the solidity of geocentric cosmology, its resistance to criticism provoked by the progress of astronomical observation, it must be noted that Aristotle's physics so harmoniously complements the spherical astronomy of Eudoxus that was refined by Ptolemy that they supported one another in a rather circular fashion (in the logical sense) until Galileo, in the early seventeenth century, dispensed with them both. Thus, in the fourteenth century in Oxford and then in Paris Aristotelian physics was, on particular points, subjected to a very penetrating critique, in which one can perceive the beginnings of a new mechanics which in certain respects anticipates modern mechanics; but these ideas were not developed, because belief in the spherical cosmology closed men's minds.

In the sixteenth century, however, Copernicus, at last reviving Aristarchus, whose work he knew well, showed that the heliocentric hypothesis gave a much more satisfying explanation of the apparent motions of the planets than Ptolemy's system, and fifty years later Kepler discovered the exact laws governing the motions of the planets around the sun.

But the Copernican hypothesis was well received only by astronomers, and Kepler never really understood the enormous importance of his own discoveries. Like his contemporaries, Kepler had never completely detached himself from the physical conceptions of antiquity; he did not realize that Galileo was engaged in founding a new physics which would be in accord with the new astronomy.

This is not the place to expound the principles of Aristotle's physics, a theory which is difficult, full of brilliant intuitions, but also naive; relatively consistent with the prescientific observation of natural phenomena, but incompatible with the results of physical experiment as we conceive it; and so alien to the elementary principles in which our scientific training has indoctrinated us, for good or for ill, that we must make a huge effort of adaptation to follow its reasoning. It is nevertheless possible, and it will be very useful for our study, to emphasize a few dominant features of Aristotle's theory of physics, which show how intimately it is related to the geocentric cosmology.

Physics, as a theory of nature, is also a theory of motion, for motion is an indisputable fact of observation, and nature is a "principle of rest and motion." The term "motion" must be understood here in a wide sense, in which it designates qualitative changes as well as displacements in space. According to this definition, the growth of a plant or the evaporation of water are just as much "motions" as the fall of a body. Let us confine ourselves to the discussion of motion in the narrow sense of the moderns, i.e., as displacement, motion "with respect to place" or "locus," in Aristotle's terminology.

Aristotle distinguishes *natural* motion and *violent* motion. Natural motions are processes by which objects, obeying a kind of internal necessity of their nature, are spontaneously carried to the place that is assigned them in the universal order. In a violent motion, on the other hand, a physical object is subjected to the coercion of a "mover," i.e., of another physical object which moves it in a direction that diverts it from its natural place. A stone falls; this is a natural motion. I throw it in the air; this is a violent motion.

There are for Aristotle only three kinds of natural motions, and their definition is closely related to the spherical and geocentric structure of the universe: rectilinear motions toward the top or the bottom; and circular motions around the center of the earth. The latter motions are the exclusive prerogative of the planets; they are without beginning or end, and performing them does not bring any more perfection to the beings that are animated in this way. On the other hand, rectilinear motions of "radial" direction, toward the top or the bottom, are natural in the sense that, in this inferior world within the orbit of the moon (the "*sublunar*" world) in which the cosmic order is constantly being disturbed, they place or replace physical objects in their natural places. The motions are completed when this end is reached; unlike the rotation of the planets, these are finite processes. This theory is closely related to the theory of the elements that make up the material world, a question which Greek rationalism had raised well before Aristotle.

The earth is not only the central sphere of the world, it is also an element, whose natural place is at the bottom. Each particle of earth spontaneously falls as close as possible to the center (indeed this is the reason which, for Aristotle—who was rather close to the moderns on this point—explains the spherical shape of the earth).* On the other hand, fire, another element, spontaneously rises; its natural place is at the top. It is true that there are two intermediary elements (Plato claimed to explain

* *Treatise on the Heavens,* 297*a–b*.

mathematically why there were two, but Aristotle does not insist on this point), water and air, whose intervention somewhat complicates the theory of place and of natural motion, since for these elements the natural motion is necessarily relative. Water, for example, falls in relation to air, but rises in relation to earth.* But after all, it is consistent with the cosmological scheme of the spheres and with ordinary observation to conceive that the natural order of the elements arranges them in concentric layers with the earth in the center, and that the relative lightness and heaviness of air and water introduce no disorder into the cosmic harmony.

The distinction between natural motion and violent motion reproduces, in the theory of motion, the fundamental distinction between the terrestrial domain and the celestial domain, and raises, in terms altogether different from those to which we are accustomed, the problem of the causality of motion. Neither the falling of bodies nor the rotation of the planets results from the action of a force. There is in the celestial domain a kind of determination of the circular motions one by another, although no one of these motions results from the action of a force acting in time (as must be the case in the causality of violent motions). The fundamental rotation, that of the eighth sphere, the sphere of the fixed stars, depends directly and as a whole on the constant action of the first mover, itself unmoving.

This is why, when the astronomer has reduced a complex celestial motion to a system of composite circular motions, he has completed his task and, in the phrase of a commentator of Aristotle, he has "saved phenomena." This is also why, when Kepler discovered, after years of observation, that Mars describes not a circle but an ellipse around the sun, this discovery plunged him into a great confusion. Although he was more of a Platonist than an Aristotelian, he remained too faithful to the conceptions of ancient cosmology to refuse the circular motion its traditional privilege.

The Aristotelian principles, when applied to the celestial

* The natural motion of the intermediary elements cannot be absolute, because there is no empty space, according to Aristotle.

motions, reduce the task of the astronomer to a cinematic analysis. To speak in this case about the "cause" of the motions is to pass beyond the realm of physics proper to another branch of philosophy. Given, moreover, that the "sublunar" world, in the cosmology of Aristotle, is not organized as rigorously as the celestial world, it is not surprising that neither Aristotle nor any physicist after him, with a few rare exceptions, brought to the description of physical phenenoma that concern for accuracy which astronomers were already using in the description of the heavens. But even as imprecise as it was in its application to particular phenomena, Aristotle's physics nevertheless encountered insurmountable problems in the description of violent motions (for example, the hurling of a projectile by a catapult) of which Aristotle and all his successors were aware. But to overcome these problems required the concepts of modern mechanics, which is incompatible with the cosmology of the sphere because it rests on the principle of inertia and it does away with the difference between celestial and terrestrial phenomena.

GEOCENTRIC ASTRONOMY

Scientific Account

I. DESCRIPTION OF THE PHENOMENA OF OBSERVATION

a1. To an observer located at any given point on the earth's surface, the earth appears, except for surface irregularities, to be a flat expanse covered by a hemispherical dome which is blue during the day and black and dotted with points of light at night. It is this dome, called the vault of heaven, or the celestial vault, which is the arena of astronomical phenomena; those stars which emit sufficient luminous energy to be perceived by the observer all seem, whatever their distances, to be part of this vault. Some of them, such as the sun, the moon, or the Andromeda nebula, have an apparent diameter which is perceptible to the naked eye; others appear as points of light which one can sometimes see in the form of small discs in a telescope (the planets); still others seem to be nothing but points no matter how powerful the instrument used (the fixed stars). However, we have no idea *a priori* of the relative distances of these different bodies. On the surface of the earth, we are familiar with the height of a man, a tree, or a house; it is easy for us to estimate the distances of objects like these and thus to have a three-dimensional view of a landscape. In the sky, however, distances are all judged to be equal and in the beginning man will concentrate his attention on studying the motions of the stars on the celestial vault; later, careful observation will lead him to suspect that these stars are either more or less remote and to try to measure their distances.

a2. The terrestrial observer located at the center of the celestial vault perceives a direction which he regards as of prime importance: the vertical. This is the direction of gravity, that is, the direction taken by falling bodies when they are allowed to drop freely, the direction of the axis of the body of a standing man, of the trunks of trees.

When a vertical line is extended upward it intersects the celestial vault at a point called the zenith. The vertical is perpendicular to a plane which corresponds roughly to a plane that seems identical to the flat earth, called the horizontal plane. The horizontal plane joins the celestial vault along the circumference of a great circle called the line of the horizon, or the horizon, which is the outer limit of what the observer can see on the surface of the earth and in the celestial vault. Finally, if one assumes that the celestial vault joins another hemispherical dome located below the horizon so as to form a sphere, the celestial sphere, the

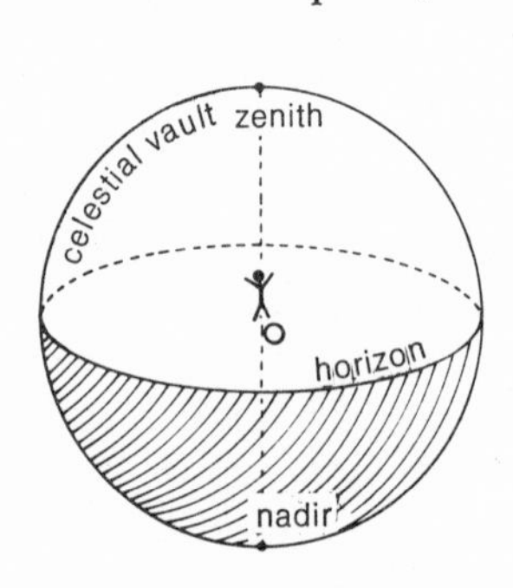

FIGURE 1

vertical also intersects this sphere below the horizon at a point diametrically opposite the zenith, the nadir (Fig. 1).

An observer who is standing up also has the ability to distinguish right from left and to picture two kinds of rotation around the vertical axis joining nadir and zenith. Rotations which draw objects along so that, when you look at them, you see them go from left to right are "clockwise," or retrograde, rotations. Rotations which, under the same conditions, draw objects from right to left are called "counterclockwise," or direct, rotations. The observer can also perceive or picture to himself direct or retrograde rotations around an axis running from the observer toward a given point on the portion of the celestial sphere above the horizon. The observer will judge in such a rotation to be direct or retrograde in relation to his body, which he will imagine to be located along the axis with the head toward the upper end.

a3. It has been known for a long time now that the earth is not flat but spherical. The portion of this sphere which an observer can see on his side of the horizon seems flat to him because of the insignificance of his height (about 1.7 meters) in relation to the radius of the earth (about 6,400,000 meters). Consequently, the horizontal plane passing through an observer *OA* is in fact the plane tangent to the earth at *O*, and the horizon is the circle of contact with the earth of the cone circumscribed from the summit *A*, if *A* is the eye of the observer. It is a commonly observable phenomenon that the horizon recedes when the observer rises above the ground. The vertical is an altogether relative direction. An object falling toward the center of the earth, on the vertical of an observer *O′A′*, is not parallel to the vertical of *OA* but meets it at the center of the earth; this is why we talk about "local vertical" (Fig. 2). The celestial sphere with its zenith, its nadir, and its horizon is also local; however, it is permissible to assume that each local sphere has as its center the center of the earth and not its respective observer, given the very small value of the radius of the earth in comparison with the distances of the stars. In the case of stars that are near, like the moon, or very near, like artificial satellites, we shall make allow-

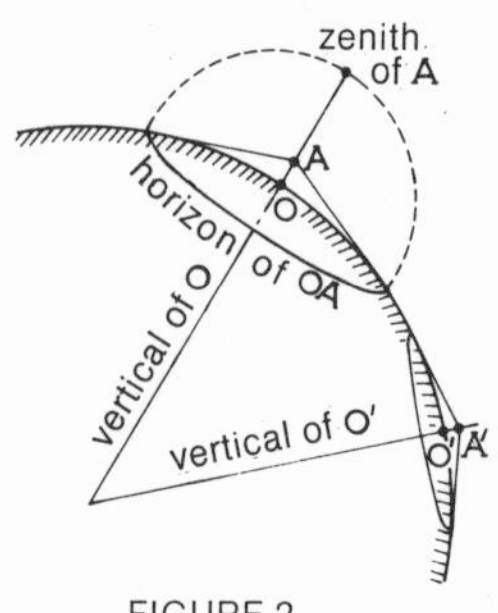

FIGURE 2

ance for corrections necessitated by what is called parallax. Similarly, it is permissible not to limit the celestial sphere to a single sphere resting on the horizon, but to picture a sphere of arbitrary radius, large in relation to the radius of the earth, in such a way that all local spheres are merged into a single and identical sphere whose center is the center of the earth. However, each observer will see only that portion of the sphere which is located above his plane of the horizon; each observer will have his own zenith, i.e., the intersection of the celestial sphere with his vertical axis.

b1. Observation of the celestial sphere at different times of the night or at different seasons of the year reveals two different kinds of stars. Certain ones, by far the more numerous, maintain fixed positions in relation to one another. They appear in the form of points of greater or lesser brilliance which the eye can group together in patterns called constellations. The imagination of the ancients associated these patterns with animals or mythological characters, whence the names of the constellations which have been preserved to this day: the Great Bear, Leo, Perseus, etc.

According to the time of day or the season of the year, the observer will see one of these constellations closer to or further from the horizon or the zenith; it may even happen that a constellation is at certain times invisible. But he will always be able to identify it on the celestial sphere by its pattern and by its position relative to neighboring constellations.

The stars which form these constellations are the ones known as the fixed stars; they seem connected in a permanent way to a sphere, called the sphere of the fixed stars, which is located just inside the local sphere and has the same radius as it, and which glides over this local sphere and creates the procession of the constellations.

The motion of the sphere of the fixed stars in relation to the local sphere is called the *diurnal motion.* It is, on the face of it, very simple: it is a movement of uniform rotation around one diameter of the local sphere. This diameter, the axis of the world, passes through the local sphere at two fixed points, the poles. One of these, the North Pole, is located in the vicinity of a star in the constellation of the Little Bear which is called, for this reason, the Pole Star (Polaris). There is no remarkable star in the vicinity of the South Pole, diametrically opposite the North Pole on the celestial sphere. Around the axis of the world, which runs from the South Pole to the North Pole, the diurnal motion occurs in a retrograde direction in a sidereal day, divided into twenty-four sidereal hours. When we say that the rotation of the sphere of the fixed stars is uniform, we are implicitly assuming that we have at our disposal a uniform scale of time in relation to which we can record the rotation of the sphere and verify that it is uniform, i.e., that the sphere turns an equal number of degrees in an equal amount of time. The question of a scale of time is a delicate one to which we shall return later on. For the moment, it is precisely this diurnal motion, assumed *a priori* to be uniform, which will define our scale of time. All speed of rotation, or all linear speed, will be measured in terms of this scale.

Incidentally, the definition of sidereal time given above is not rigorously accurate, as we shall see further on.

b2. For an observer who has the North Pole above his horizon, the earth's diurnal motion causes the stars to appear above the horizon and to rise until they "culminate" at a point on the local "meridian," a great circle formed by the intersection of the local sphere with the plane defined by the earth's axis and the local vertical (the meridian plane). Afterward, the stars fall again. Some fall far enough so that they set beneath the horizon. The meridian plane cuts the horizon at two diametrically opposite points, North and South. The diameter of the horizon, which is perpendicular to the line North-South, meets the horizon at points East and West, as is indicated on the drawing (Fig. 3).

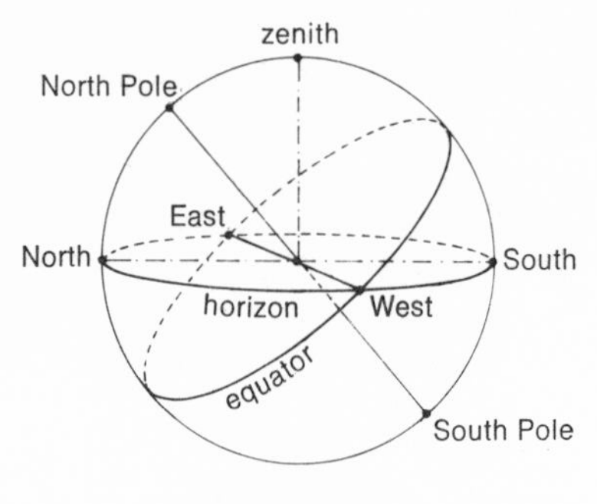

FIGURE 3

Finally, the great circle of the sphere of the fixed stars whose plane is perpendicular to the earth's axis is called the *celestial equator;* it meets the horizon at points East and West.

The diurnal motion may be regarded as the real motion of the sphere of the fixed stars around the earth's axis or, on the contrary, as an apparent motion caused by the fact that the earth is turning in a counterclockwise direction on its axis. It is quite obvious that either hypothesis explains perfectly what is seen by an observer of the celestial vault. The earth's axis passes through the earth at two points which are called the North and South Poles. The plane of the celestial equator intersects the earth along a great circle which is the terrestrial equator. It is easy to see that, for an observer located at the earth's North Pole, the plane of the horizon merges with the plane of the celestial equator, the zenith coincides with the celestial North Pole, and the diurnal motion draws each star in a retrograde direction along a small circle parallel to the horizon. No star rises or sets, and the only stars visible are those located in the northern hemisphere of the sphere of fixed stars, that is, in the hemisphere which contains the North Pole. For an observer located at the South Pole, the situation is analogous, but each star describes a circle parallel to the equator in a motion which is direct in relation to a vertical line running from the observer to the zenith; the only stars visible are those in the southern hemisphere of the sphere of the fixed stars (Figs. 4, 5, and 6).

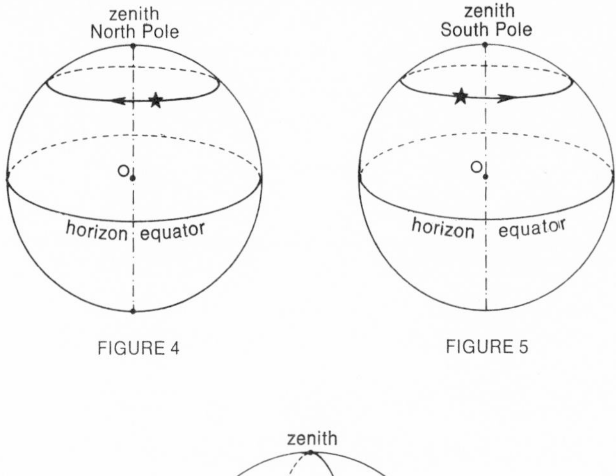

FIGURE 4

FIGURE 5

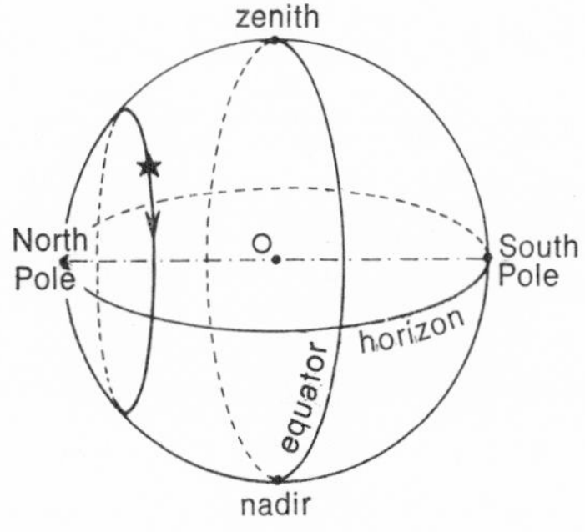

FIGURE 6

Let us assume that we are standing at a point along the terrestrial equator. The earth's axis is then contained within the plane of the horizon and the stars describe circles which are perpendicular to this plane. All stars rise and set and all the stars belonging to the sphere of the fixed stars pass before the observer. If the observer is located at an intermediary position on the earth, for example at a point located between the equator and

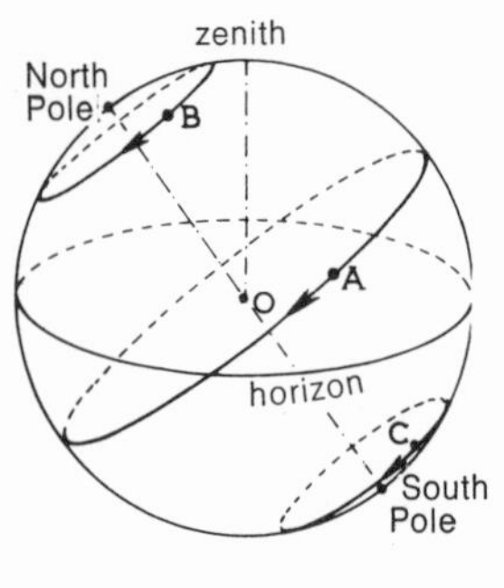

FIGURE 7

the North Pole, one can easily see on the drawing that star *A* rises and sets (Fig. 7). Star *B* remains constantly above the horizon and twice a day passes through the meridian above the horizon; such a star is said to be circumpolar. Star *C* will never be visible.

c1. There are certain stars which, although they are also involved in the diurnal motion, have no fixed positions among the constellations but occupy different constellations in succession. These are the sun, the moon, the planets, and the comets.

When the sun is above the horizon of a given place, we say that it is day in this place; when it is below the horizon, we say that it is night. The enormous amount of light energy received from the sun is diffused by the earth's atmosphere, which then appears to be blue, and prevents one from seeing those stars which are above the horizon. It seems as if it would be difficult to determine whether the sun is also a fixed star or whether it moves from one constellation to the next. However, one can see those stars which are in the vicinity of the sun just before it rises or just after it sets. If the sun were a fixed star, it would always be the same stars that culminated at the meridian at midnight, that is, twelve hours after noon, the time when the sun passes through the meridian. These stars would be the ones diametrically opposite the sun on the sphere of the fixed stars. But this is by no means the case: at the beginning of March, for example, the constellation Leo reaches the meridian around midnight; at the end of May, it is the constellation Scorpio.

It is actually a simple matter to transfer to a globe representing the sphere of the fixed stars the different positions of the sun at different times of the year. Thus one observes that in a year of approximately 365 days, the sun describes by direct motion a great circle of the sphere of the fixed stars, called the ecliptic. This great circle intersects the celestial equator at two points, the nodes. The sun passes through one of these nodes at the end of March when it passes from the southern hemisphere to the northern hemisphere. The node corresponding to γ is called the ascending node, the equinoctial point, or the vernal (or spring) equinox. At the end of September the sun passes from the northern hemisphere to the southern hemisphere by crossing the equator at the descending node, or autumnal equinox. The plane of the equator and the plane of the ecliptic form an angle of approximately 23° 30′ (Fig. 8).

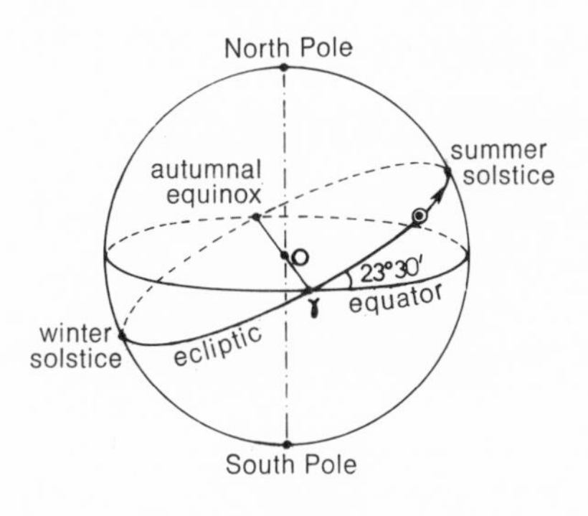

FIGURE 8

c2. The equinoxes are not absolutely fixed on the sphere of the fixed stars but are moving very slowly in a retrograde direction. This motion is called the *precession of the equinoxes,* and it can be described briefly as follows. The ecliptic is fixed in relation to the stars; the vernal equinox is moving along the ecliptic in a retrograde direction and travels all the way around the sphere of the fixed stars in approximately 26,000 years. As a consequence the equator, which forms a fixed angle with the ecliptic, is not absolutely fixed in relation to the stars. Similarly, the line connecting the poles, or the earth's axis, which is perpendicular to the equator, is not fixed either, but describes in 26,000 years a cone whose axis is a line perpendicular to the plane of the ecliptic leading toward the center of the sphere of the fixed stars and intersecting this sphere at two points, the north and south poles of the ecliptic. The half-angle at the summit of the cone equals 23° 30′, and thus the North Pole of the earth's axis is not fixed in relation to the stars, but moves along a small circle. Our present pole star will eventually lose its role and in a few thousand years may be replaced by Vega. The phenomenon of precession is, as we see, very slow; rudimentary observation, even when continued over several centuries, is not sufficient to make it evident. Thus the North Pole appears to be fixed once and for all.

Let us now return to the definition of sidereal time. A sidereal day, divided into twenty-four sidereal hours, is the interval between two successive transits of the vernal equinox over the meridian of a given place. Since the vernal equinox is shifting slowly in relation to the stars, this interval is not the length of time it takes for the same star to reappear at the meridian; it is slightly shorter, but by a very narrow margin.

At the present time the vernal equinox is located at the boundary between the constellation Aquarius and that of Pisces and the autumnal equinox is located at the boundary between Leo and Virgo.

c3. The sun's motion along the ecliptic does not take place at a uniform speed. It has long been known that at certain times of the year (around the second of January) this motion is more rapid than at others. Let us picture an imaginary star, the "average" sun, which travels around the equator in one year—the time it takes the real sun to move around the ecliptic—at a uniform speed. We shall find that if the real sun and the average sun start together on the second of January, the real sun will get ahead of the average sun and then gradually slow down until six months later it is passed by the average sun, which it will catch up with again the following second of January. On the second of January the sun is said to be at perigee (literally in Greek, "close to the earth"); six months later, it is at apogee (Greek for "far from the earth").

c4. Study of the moon's motion against the sphere of the fixed stars is facilitated by the fact that its light does not prevent us from seeing the stars. This motion is rapid; in an hour or two of careful observation, one can see very clearly that the moon has moved in relation to the stars. If, as was done with the sun, one takes note of the successive positions which the moon occupies on the sphere of the fixed stars, one finds that it describes, by direct motion, a great circle of this sphere. The plane of this great circle forms an angle of approximately 5°9′ with the plane of the ecliptic, and the great circle itself intersects the ecliptic at two points, the ascending, or north, node and the descending, or south, node. The moon crosses the ecliptic at the ascending node when it passes into the northern hemisphere, and crosses the ecliptic at the descending node when it passes into the southern hemisphere. The moon's orbit among the fixed stars is far from being fixed. In particular, the nodes are moving in a retrograde direction and move all the way around the ecliptic in 18⅔ years; also, the tilt of the plane of the moon's orbit in relation to the ecliptic varies by 18′ in 173 days. All these phenomena were well known to Tycho Brahe in the early sixteenth century.

The interval between two appearances of the moon in the vicinity of the same star is called the moon's sidereal cycle, and is equal to approximately 27 days 7 hours 43 minutes. Since the sun is also moving in a counterclockwise direction against the sphere of the fixed stars, the moon will be in conjunction with it about every 29½ days (29 days 12 hours 44 minutes, on the average). This period is known as the moon's synodic cycle, or the lunar month. It may vary as much as thirteen hours from its average length because of the fact that the moon's motion against the sphere of the fixed stars is no more uniform than the sun's. The moon's motion is more rapid at a point called the perigee of the lunar orbit, and it slows down fifteen days later at a point called the apogee of the lunar orbit. Perigee and apogee are not fixed either, but make a complete revolution of the celestial sphere in eight years, 311 days.

The moon's synodic cycle is related to a very important and

very obvious phenomenon, the phases, which we shall describe further on.

c5. There are five planets visible to the naked eye: Mercury, Venus, Mars, Jupiter, and Saturn. They appear as luminous points of some brilliance and can be distinguished from the stars only by their rapid movements against the sphere of the fixed stars. These movements are very complex, but one can divide the planets into two groups: Mercury and Venus belong to the first group, the so-called inferior planets, and the others are known as the superior planets.

Mercury and Venus always remain in the vicinity of the sun. Sometimes they are found to the west of the sun, in which case they are visible in the morning before sunrise, and sometimes to the east of the sun, when they are visible in the evening after sunset. These two planets seem to sway back and forth from one side of the sun to the other, in a period of 116 days for Mercury, and of 584 days for Venus. These periods are known as their synodic cycles. The angle formed by the two lines earth-sun and earth-planet is known as the planet's elongation. The elongation of Mercury reaches a maximum of 24°, that of Venus approximately 44°. When a planet's elongation is zero, we say that the planet is in conjunction with the sun. In the course of one synodic cycle, each of the inferior planets forms two conjunctions with the sun. One of these occurs when the planet is moving from west to east against the celestial sphere, and is called an inferior conjunction; the other, when the planet is moving from east to west, is called a superior conjunction.

The movements of the planets Mars, Jupiter, and Saturn are altogether different; their elongations may be any number of degrees between 0° and 180°. When its elongation is zero, the planet is in conjunction with the sun; when it is 180°, it is in opposition. The planet is in quadrature when its elongation is equal to 90°.

In their complex movements against the sphere of fixed stars, all the planets move now in direct motion, now in retrograde motion, but the time spent in direct motion is longer, so that eventually each planet makes a complete revolution of the sky. Their meandering paths are very close to the ecliptic and are located within a band, called the zodiac, defined by two circles of the celestial sphere which are symmetrical in relation to the ecliptic.

d1. Up to this point, none of the phenomena studied has allowed us to imagine that the stars are not all the same distance away from the observer. It would be perfectly permissible to regard the sphere of the fixed stars as a material reality, a globe to which the fixed stars are attached, and the wandering stars, sun, moon, and planets as moving against this sphere.

The phenomena which we are now going to describe force us to abandon this hypothesis and to conclude that the stars are located at very different distances. The celestial sphere nonetheless remains a convenient instrument with which to study the movements of the stars as long as we confine ourselves to studying the movement of a straight line running from the eye of the observer to a given star.

A first phenomenon which permits us to assume that the moon is closer than any other body is the phenomenon of occultation. Stars located near the eastern edge of the moon disappear and then reappear at the western edge after the moon has changed its position in the course of its movement across the sphere of the fixed stars.

The phenomenon of the phases of the moon is quite spectacular. Whenever the moon reaches the meridian (about twelve hours after the sun, that is, at midnight), it appears in the form of a circular disc that is completely illuminated: the moon is full. As the moon follows its monthly course toward the east, the portion of the disc that is illuminated gradually becomes smaller. At the end of a week, we can see only the half of the disc that is turned toward the sun, called the last quarter; then the moon takes on the appearance of a crescent that becomes thinner and thinner and disappears one lunar half-month after the full moon: now the moon is "new" (Fig. 9). Then it reappears to the east of the sun, in the shape of a crescent; the portion that is illuminated increases every day until the full moon, after going through the intermediary stage of the first quarter.

This phenomenon of the phases can be explained very satis-

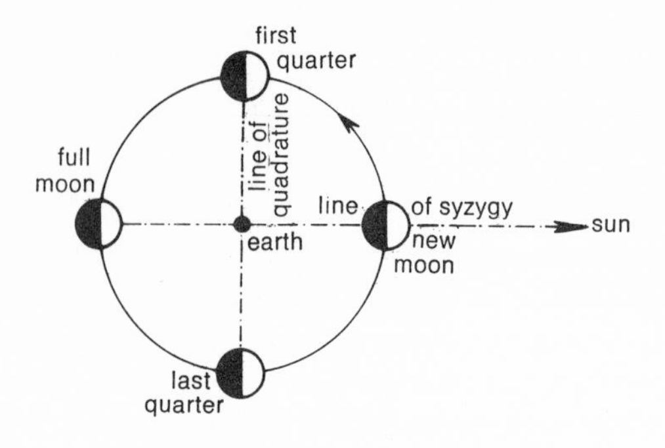

FIGURE 9

factorily by assuming that the moon is a sphere which has no light of its own and one-half of which is illuminated by the sun. According to the angle formed by two straight lines, one passing from the earth to the sun and one from the earth to the moon, the illuminated portion of the moon that is visible from the earth varies (Fig. 10). At the time of the new moon, the sun and the moon are said to be in conjunction. They are in opposition at the full moon, and in quadrature at the first and last quarters.

d2. Let us assume that the moon's orbit is in the plane of the ecliptic. At each full moon the centers of the sun, the earth, and the moon would be aligned, and since the earth would have come between the rays of the sun and the moon, the moon would cease to be illuminated and would disappear. There would be an

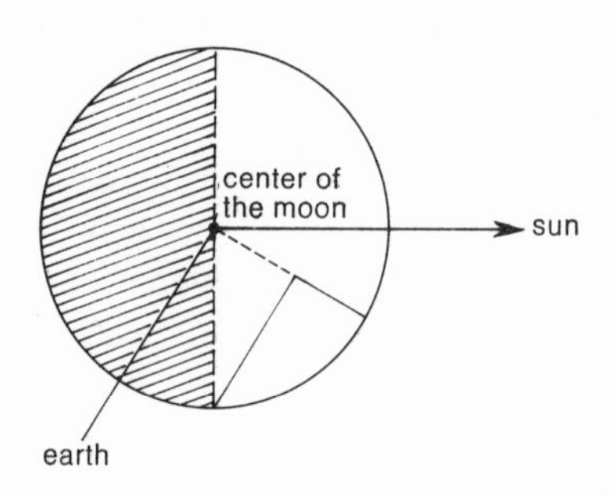

FIGURE 10

eclipse of the moon. At the new moon, on the other hand, the moon, passing in front of the disc of the sun, would screen it from the sight of some observers on earth, since it happens that the apparent diameter of the moon is equal to that of the sun (around 32′ of arc). There would be an eclipse of the sun.

We have seen that, in fact, the moon's orbit is tilted in relation to the ecliptic by an angle of approximately 5°, with the result that, generally speaking, there is no eclipse at the time of full moons or new moons. For there to be an eclipse, it is necessary not only to have a conjunction or an opposition, but also that the moon be close to the ecliptic, that is, in the vicinity of one of its nodes. It happens that 223 lunar months have a duration approximately equal to 242 "draconic" months. A draconic month is the interval that separates two transits of the moon over its ascending node, taking into account the retrograde motion of this node along the ecliptic to which we have referred. As a consequence, at the end of this interval which includes 223 lunations or 242 draconic months, eclipses recur in the same order as previously. This cycle, called Saros, makes it possible to predict eclipses with a fair degree of accuracy, but of course, if one wishes to be very precise about the time and circumstances of an eclipse, one must make complex calculations which take into account all the formulas which translate the movements of the sun and moon against the celestial sphere.

Eclipses of the moon are visible from one entire hemisphere of the earth, the hemisphere that is turned toward the moon, since this phenomenon is caused by the fact that the moon is no longer being illuminated. Eclipses of the sun, on the other hand, are visible only from certain places, those places located within the shadow cast by the moon. Because of the earth's rotation on its axis, the places from which one can see a total eclipse of the sun form a band whose width is approximately 200 kilometers. On either side of this band the eclipse is partial, which means that a portion of the solar disc remains visible. Further away from the band of total eclipse, there is no eclipse at all.

d3. There is no phenomenon clearly perceptible to the naked eye that enables us to state that the planets are dark spheres illuminated by the sun whose distance from the earth varies. However, with a telescope of modest dimensions one can observe that these planets have the appearance of discs and that in addition they have obvious phases, exactly like the moon. The phases of Venus are easily visible and were discovered by Galileo. At certain times Mercury and Venus pass in front of the disc of the sun and can be clearly distinguished as black dots on its surface.

As for the stars, while even in the most powerful telescopes they can be seen only as points, one begins to suspect that their distances from the earth are not equal by observing the parallaxes of the ones closest to earth. Let us assume that a number of stars, *B*, *C*, *D*, and *E*, are extremely remote and that another star, *A*, is relatively close, and that *A* appears to be close to two of the first group, *B* and *C*, for example (Fig. 11). If the earth moves through space so that after starting from *T′* it arrives at *T*, star

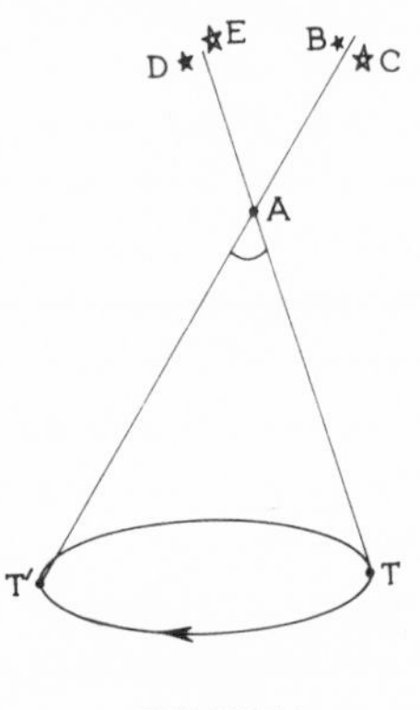

FIGURE 11

A will appear to be close to stars *D* and *E* rather than to stars *B* and *C*, as can be seen on the drawing. The angle *TAT′*, in which *T* and *T′* are the positions occupied by the earth at an interval of six months, is called the parallax of star *A*. This angle is very much exaggerated in the drawing; it never exceeds 1″ for any star, which makes its measurement very difficult.

The parallax of the moon is very easy to measure. The base utilized is not formed by two points occupied by the earth at an interval of six months, but by two points very far apart located on the surface of the earth. If the two points in question are diametrically opposite on the earth, the distance between them is about 12,800 kilometers and the parallax of the moon is then equal to 2°. This parallax is easily measurable and makes it possible to calculate the distance from the earth to the moon (approximately 378,000 kilometers—or 236,000 miles—on average).

II. LOCATION OF PHENOMENA

Let us return to the diurnal motion. It is obvious that one can interpret the rotation of the sphere of the fixed stars as an appearance caused by the rotation of the earth. The diurnal motion plays a very important part in human life, if only because it periodically brings the sun back above the horizon of a given place; and it makes very little difference what is really turning, since the effects are exactly the same whether it is the sphere of the fixed stars or the earth. It is actually more convenient to assume that it is the celestial sphere that is turning, for we have no sensation at all of the earth's movement, but we do see the stars moving in procession before our eyes. This is why the systems of coordinates we are going to describe, which are all geocentric, have not only been utilized for a very long time, but always will be. It is not until we wish to explain certain phenomena less important than the diurnal motion, like the complex movements of the planets, that we shall have to devise other systems of coordinates which are nongeocentric, and whose purpose will be precisely to render the appearances of phenomena as simple as possible.

a1. HORIZON COORDINATES. Let us assume that, at a given moment, star *A* is on the local sphere of observer *O* (Fig. 12). Line *OA* and the plane of the horizon form an angle *h* which is known as the altitude of the star. The plane containing *O*'s vertical and star *A* and the meridian plane form a dihedral angle *a*, which is known as the azimuth of the star. The altitude is measured from 0° to 90° from the horizon to the zenith, and from 0° to −90° from the horizon to the nadir. The azimuth is measured from 0°

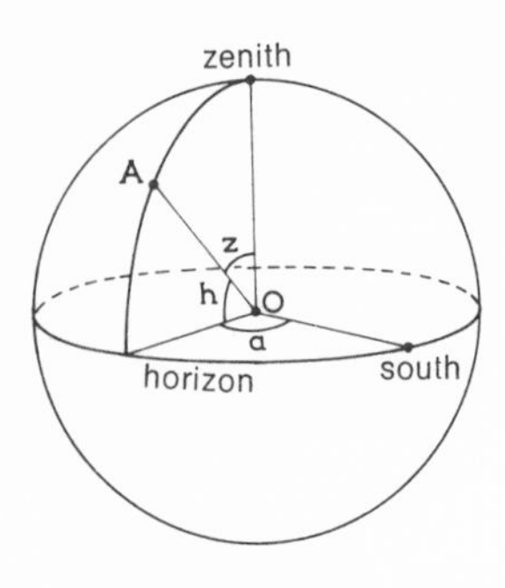

FIGURE 12

to 360° in a clockwise direction starting from the southpoint of the horizon. Angle *z* formed by line *OA* and the line of the zenith (or ascending vertical) is the complement of the altitude; it is known as the zenith distance, and it varies from 0° to 180° between the zenith and the nadir.

a2. HOUR COORDINATES. Let us now consider the great half-circle of the celestial sphere that passes through the line of the poles and through star *A* (Fig. 13). This great half-circle meets the equator at point *B;* it is called the hour circle of star *A*. Angle *d* between *OB* and *OA* is called A's declination. It is measured from 0° to 90° from the equator to the North Pole, and from 0° to −90° from the equator to the South Pole. The plane of *A*'s hour circle and the meridian plane form a dihedral angle *H* which

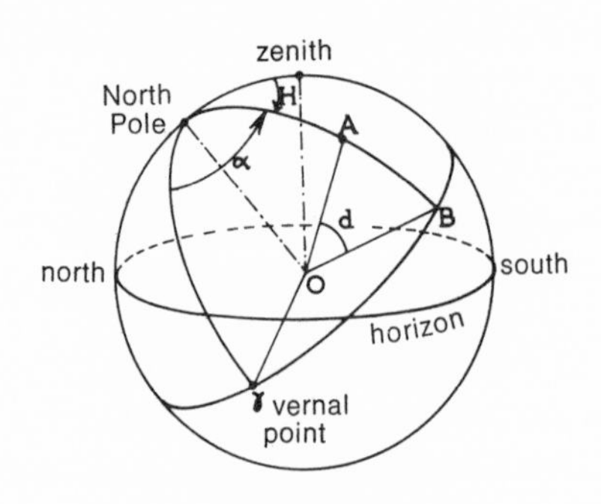

FIGURE 13

is called the hour angle of A. This angle is measured in hours from 0 to 24 in a clockwise direction starting from the southern half-meridian, in accordance with the ancient custom of measuring angles in terms of a unit called an hour (24 hours being the equivalent of 360°). The confusion is increased by the fact that, if the star in question is at the vernal point, its hour angle is given the special name "sidereal time," but if one is talking about the center of the sun, its hour angle is given the name "true solar time," whereas what one is really talking about is angles. We shall return to these questions.

a3. EQUATORIAL COORDINATES. The dihedral angle formed by the plane of the hour circle of the vernal point γ and the plane of the hour circle of A is known as the right ascension of A (see Fig. 13). This angle α is measured from 0 to 24 hours in a counterclockwise direction. At a given moment, the sidereal time at 0 is determined by reference to the hour angle of the vernal point in that place at that moment (represented by T); then we have the following formula, using the definitions we have agreed on:

$$\alpha = \mathrm{T} - \mathrm{H}$$

If star A passes through the meridian, its hour angle (H in the formula) is zero; therefore, sidereal time in a given place at a given moment is equal to the right ascension of those stars which are passing through the meridian of the place at that moment. Knowing the right ascension and declination of a star enables one to locate this star precisely on the sphere of the fixed stars. In the case of a fixed star these coordinates, known as equatorial coordinates, are fixed, provided one overlooks the gradual shifting of the vernal point caused by the precession of the equinoxes. Since this shifting is known, it will always be possible to know, for a specific instant of time, the equatorial coordinates of a star.

a4. ECLIPTIC COORDINATES. The angle formed by line OA and the plane of the ecliptic is called the latitude of A. It is measured from 0° to 90° from the ecliptic toward the North Pole of the ecliptic, and from 0° to −90° from the ecliptic toward the South

Pole of the ecliptic. The dihedral angle formed by the half planes passing through the axis of the ecliptic and through the vernal point and star *A* respectively is called the longitude of *A*. It is measured from 0° to 360° in a counterclockwise direction starting from the vernal point (Fig. 14). Since the ecliptic is by definition the path of the sun, we see that the latitude of the sun is always zero, and that its longitude increases by 360° in a year. The longitude and latitude of a star, like its right ascension and declination, enable one to determine the position of that star on the sphere of the fixed stars. It is more convenient to utilize ecliptic coordinates than equatorial coordinates to study the movements of the moon and the planets, since the latitude of these bodies remains small and their paths are close to that of the sun. On the other hand, the system of equatorial coordinates is related to the movement of the earth on its axis and is convenient to utilize for locating fixed stars in the field of a telescope. If the telescope is pointed toward a star and if it is rotated at a uniform speed around a fixed axis parallel to the earth's axis by 360° in 24 hours, the star will remain constantly in the field of the telescope from the time it rises to the time it sets.

a5. Given the ecliptic coordinates of a star, one can work out its equatorial coordinates by employing the formulas of spherical trigonometry. One takes as a unit of measurement the radius of the celestial sphere and on the surface of this sphere one considers "triangles" whose sides are arcs of great circles belonging

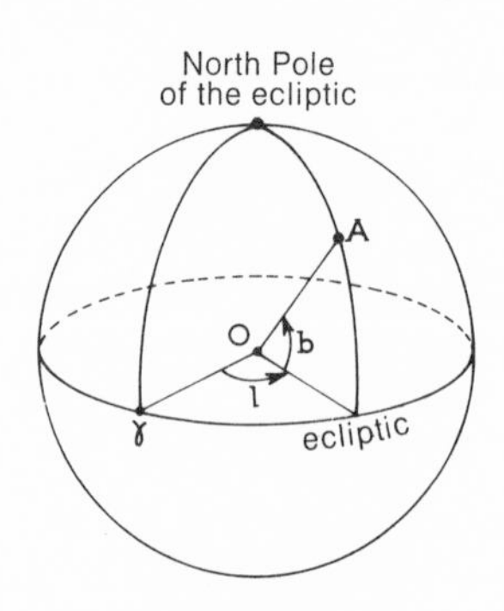

FIGURE 14

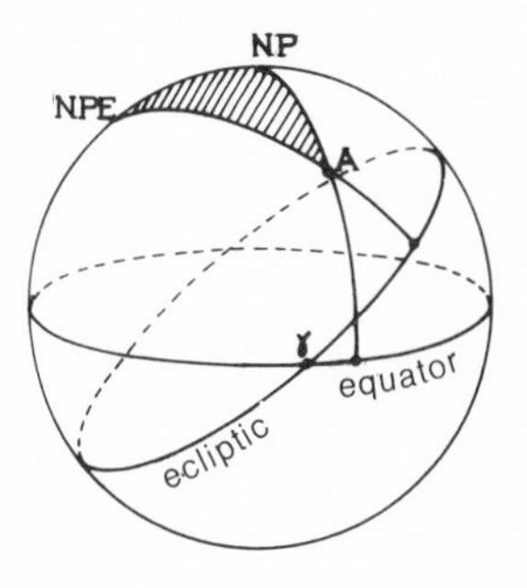

FIGURE 15

to this sphere (Fig. 15). One can then work out a group of formulas connecting the cosines or sines of the arc sides with the cosines or sines of the dihedral angles formed by the planes of the sides of a triangle in relation to a summit. For example, in the drawing, if A is a star, NP the North Pole, and NPE the North Pole of the ecliptic, the sides of the triangle are:

NPE–$A = 90° - b$, if b is the latitude of A.
NP–$A = 90° - d$, if d is the declination of A.
NPE–$NP = i$, i being the inclination of the equator to the ecliptic, or 23° 30′.

The angles are:

at NPE, $90° - l$, if l is the longitude of A.
at NP, $90° + \alpha$, if α is the right ascension of A.

One then arrives at the following formula, among others:

$$\sin b = \cos i \sin d - \sin i \cos d \sin \alpha$$

If star A is the sun, b is zero and we have

$$\sin \alpha = \text{cotan g } i \text{ tan g } d,$$

a formula which shows that one need only know the declination of the sun in order to determine its right ascension. This formula reflects the fact that the sun is the fundamental reference in determining right ascension, which is not surprising, since the vernal point is defined in terms of the path of the sun.

b1. As we have said, the regular and indefinitely repeated rotation of the sphere of the fixed stars gives concrete form to the idea of a scale of time. If we overlook the phenomenon of precession, the hour angle of the vernal point—which periodically increases by 360° (or 24 hours)—may be taken as a definition of time; which explains why this angle has been given the name sidereal time. A sidereal day, the period of time that separates two passages of the vernal point over the meridian, is divided into twenty-four sidereal hours, each hour into sixty sidereal minutes, each minute into sixty seconds.

To locate a phenomenon of daily life on the sidereal scale of time is not very convenient, for what regulates the course of human life is not the return of the vernal point or of a given star to the meridian, but rather the return of the sun, which causes daytime to follow nighttime.

Let us use the term "true solar time" to refer to the hour angle of the sun. This angle varies from 0 to 24 hours in a period of time which we shall call the "true solar day." It is apparent that the true solar day will be longer than the sidereal day, for since the sun moves in a counterclockwise direction on the sphere of the fixed stars, it will fall behind the vernal point.

True solar time, although convenient, has the serious disadvantage of not being uniform, that is, of not being a linear function of sidereal time, and is not reducible to

$$T_s = AT + B$$

if A and B are constants, T is sidereal time, and T_s true solar time. In fact, the second half of the above equation would have to be corrected by accounting for certain periodic functions, called inequalities, which are of two kinds. Some inequalities are caused by the fact that the sun's movement along the ecliptic is more rapid at its perigee than at its apogee; these are called "equations of the center." Other inequalities are caused by the fact that the ecliptic is tilted in relation to the equator; these are called "reductions to the equator." The sum of the equation of the center and the reduction to the equator is known as the "equation of time." By correcting true solar time to allow for the equation of time, we obtain a linear function of T:

$$T_m = AT + B$$

which is called "average solar time." It is this mean solar time which is given by our watches; it is equal to the hour angle of the mean sun, which we have already defined.

b2. Naturally, all the times we have considered up to now are local times, which means that they are not identical at the same moment for two separate points. The difference between two times of the same kind (average solar times, for example) in two places at a given moment is equal to the dihedral angle measured in hours (360° being equal to 24 hours) between the meridian planes of the two places. If the two places are A and Greenwich, the difference in their longitudes is known as the longitude of A, the longitude of Greenwich being 0. Longitude is measured from 0 to 24 hours from west to east. If, in addition to the longitude of a place, one also knows its latitude (i.e., the angular altitude of the North Pole over the horizon of the place), one has defined its geographical coordinates, longitude and latitude, which should not be confused with ecliptical coordinates. Geographical coordinates enable one to determine a point on the surface of the earth. They enable one to translate the coordinates of a star on the sphere of the fixed stars (equatorial or ecliptical coordinates) into local coordinates (horizontal or hour coordinates), provided one knows the sidereal time.

b3. In dealing with long periods of time, it is useful to have a unit of time longer than the day. The year, which is related to the sun's motion against the sphere of the fixed stars, is a convenient unit. A distinction is made between the sidereal year, the period of time that passes between two appearances of the sun in the vicinity of the same star, and the tropical year, the period of time that passes between two transits of the sun over the vernal point. It is the tropical year that plays an essential role in human societies, since it governs the return of the seasons. Spring is the period of time between the sun's crossing of the vernal point and the moment when its longitude is equal to 90°. At this moment the sun's declination is equal to 23°30′; this declination is passing through its maximum and hence varies very slowly: this is the summer solstice. Summer is the period of time between the summer solstice and the sun's transit over its descending node, or the autumnal equinox; autumn is the period of time between the autumnal equinox and the moment when the sun's longitude is equal to 270°; its declination is then equal to −23°30′, it is passing through its minimum and varies very slowly: this is the winter solstice. Finally, winter is the period of time when the sun passes from the winter solstice to the vernal point.

III. ELEMENTARY NOTIONS ABOUT THE PHYSICS OF THE STARS

Although up to now we have considered the stars merely as points or as moving discs, we can now make certain statements about their possible physical nature, statements which are suggested by some extremely simple observations.

The stars do not have the same degree of brilliance, or "luminosity." The most brilliant ones are said to be of the first magnitude; the weakest ones which are still visible to the naked eye are said to be of the sixth magnitude; the rest are assigned magnitudes between 1 and 6. This notion of magnitude has a precise mathematical basis; it is a linear function of the logarithm of the degree of luminosity produced by the star. Magnitude is defined in such a way that the dimmer the star, the higher its magnitude. Of course, magnitude is only apparent, for two identical stars may have different degrees of luminosity if their distances are different. Measuring the distances of the stars by measuring their parallaxes, for example, in the case of the nearest stars, enables us to establish an absolute magnitude, i.e., the magnitude a star would have if it were located at a given distance from the observer. From this one observes that the stars do not all have the same intrinsic luminosity, certain bodies being much more brilliant than others. Another phenomenon of observation that is easily accessible is the fact that the stars are not all the same color. Rigel is blue, Aldebaran red, Vega white, the sun yellow. The precisely quantified evaluation of these colors by means of a "color index" results in the classification of the stars in color categories designated by the letters O, B, A, F, G, K, and M. Blue stars belong to class O, red stars to class M, yellow stars to class G, etc. An experiment frequently conducted on the surface of the earth—heating a metallic body until it becomes red hot or white hot—suggests that the color category of a star corresponds to its temperature: O stars are the bluest and therefore the hottest, and M stars are the reddest and therefore the coldest.

As for the moon and the planets, the small amount of light that we receive from them, considering their relative proximity, the phenomenon of phases, and the qualitative similarity between the light they send us and the light of the sun suggest that here we are dealing with cold bodies which are merely reflecting the light of the sun without emitting any light of their own.

PART TWO

CLASSICAL SCIENCE AND THE DECLINE OF THE UNIVERSE

Philosophical Account

The collapse of the ancient cosmology began in 1543 with the publication of Copernicus' famous work, *De revolutionibus orbium coelestium, libri VI* (*Six Books on the Revolutions of the Celestial Orbs*). But this was only a beginning, and the same reasons we have offered for the remarkable stability of the ancient cosmology also help to explain why the Copernican revolution, which displaced the earth and put the sun in the center of the world, was in reality only the opening shot of the cosmology revolution and why, almost a hundred years after Copernicus, it was still possible for a court of law to condemn Galileo to death. A counterrevolutionary act that failed, to be sure, but one that was enough to prove that a mere alteration, however radical, in the way of describing the celestial motions was not sufficient to overthrow the ancient cosmology. It was also necessary to eradicate Aristotle's physics and, with the aid of Archimedes' lever, to make the world rotate in infinite space: it was necessary to have Bruno, Galileo, Descartes, Newton.* It was no accident,

* For Copernicus, just as for Eudoxus, Aristotle, and Ptolemy, the universe is a system of concentric spheres in which the sphere of the fixed stars contains all the others; but it is the sun rather than the earth that is in the

however, that the astronomical superstructure was the first to give way. It was, in fact, much more difficult to refute Aristotle's physics, in spite of its obscurities and inconsistencies, which had long been known, than Ptolemy's astronomy for the simple reason that the second was much closer to a modern scientific theory than the first. Ptolemy's system could be subjected much more easily than Aristotle's physics to the criterion which Karl Popper and today's epistemologists generally regard as characteristic of a real theory in the scientific sense of the word: it could be disproved and refuted by means of observation and experiment.* Precisely because it was geometric, precise, and truly "scientific" in the modern sense, the Ptolemaic interpretation of the celestial motions was *capable* of being refuted by observation, much more so than the physics of Aristotle, which was not geometric, precise, or "scientific."

There would be no point in enumerating the mass of facts which, reflecting the advances in astronomical observation (very significant by the sixteenth century and enormous after Galileo), was destined to render Ptolemy's theory more and more doubtful and to increase what we would now call the credibility of the Copernican hypothesis.

Furthermore, the idea of a difference between celestial and terrestrial phenomena became more and more untenable. It was Galileo who, thanks to his famous telescope, furnished the decisive proof on this subject by demonstrating in a manner which three hundred years afterward remains exemplary, that the sun has spots and that the moon, like the earth, is covered with mountains. Before him, Tycho Brahe and Kepler had carefully observed novae, stars which suddenly glow with a brief but dazzling brilliance and then fade, thus refuting the old dogma of the permanence and incorruptibility of celestial objects. Against this dogma, Bruno, a less good observer and a mediocre mathematician

---

center of the system. The radius of the sphere of the fixed stars is incommensurable (*incomparabilem*) with the distance that separates the earth from the center of the sun. The point is not that this radius is infinite, but that we cannot measure it by comparison with sizes that are empirically accessible.

* And of course, it could also be confirmed by experiment, but Popper has shown that, whereas a theoretical statement about the world can be absolutely and definitively refuted by experiment, no such statement can ever be absolutely and definitively confirmed by experiment.

but a subtle and profound dialectician, had advanced the argument which is most impressive, at least to the moderns, and which may be the most decisive: the very fact that the stars shine, he said in less simple language, proves that they are not incorruptible.

But the discovery that celestial phenomena are not essentially different from terrestrial phenomena sent scientists back from astronomy to physics. They needed a theory of natural phenomena which supported this identification, or at least this homogenization, but it was not until Newton that this theory was finally framed. Thus it is not until the end of the seventeenth century that one can talk about a new cosmology, which we have agreed to call "classical" to distinguish it from the two others, the ancient and the modern.

Actually, if one examines history closely, one is inclined to be somewhat skeptical about the very idea of a "classical cosmology," for some important reasons which we have cited ourselves in our introduction and which have to do with the actual development of the philosophy of nature.

Until the middle of the eighteenth century, two systems of the world were in competition as to which would succeed the cosmology of Aristotle—Descartes's and Newton's. In fact, this implies a false symmetry, for while Descartes had explicitly constructed a cosmology, Newton had taken care, according to his famous maxim, not to invent *hypotheses*, but had confined himself to stating, on the basis of his observations, a few absolutely universal facts—the absolute character of space and time (I, II), the laws of mechanics (III), the law of universal gravitation (IIIb5)—and had left to others the task of creating a world out of the elements of this "natural philosophy."

But for those living at the time of the celebrated and empty disputes which set the Cartesians against the Newtonians, it was indeed two systems of the world that were in conflict, and the English cosmology (which was actually supported by an illustrious Frenchman, Voltaire) was no less incompatible with the French cosmology (which in fact numbered some respectable advocates in England) than either was with the Greek (which nobody defended any more).

The ultimate triumph of Newton's system at the end of the

eighteenth century is really the triumph of a theory of physics, in the modern sense of the term, rather than the triumph of a cosmology. For this triumph casts back into the darkness certain questions that are fundamental from the point of view of cosmology: the questions of absolute time and space (already raised by Leibnitz), the meaning of the principle of inertia (raised by Berkeley), and the possibility of action at a distance (a battle horse which the Cartesians rode to death). These questions were all destined to reappear in a new light, a century later, with the dawn of modern physics. But while waiting for this resurgence, and at the time that these questions were being forgotten, it was, as we have said, cosmological thought itself that was being eclipsed. Are we to conclude that between the general defeat of the Cartesian system and the beginnings of modern cosmology, cosmology itself no longer existed? In other words, that apart from the Cartesian episode, which was short-lived, and a few more or less sporadic attempts like that of Kant, there was no such thing as a classical cosmology?

The answer is yes, in a sense. For there was, after the eighteenth century, no theory of the universe, in the full meaning of the word, that was explicitly built on the foundation of classical science, systematically defended and developed, and then tested against observation. The fact that Laplace's "system of the world" is actually a theory of the solar system (Va), when Laplace and all his informed contemporaries knew perfectly well that the solar system is surely only a very small part of the universe—this fact is, in this connection, very significant. Laplace's "world" is that part of the world about which he thinks he is able to reason, on the basis of well-established phenomena of observation and with the aid of a sound theory of physics (in point of fact, Newton's mechanics and the law of universal gravitation). Laplace constructs no hypothesis about the properties of the universe as a whole, except that in it the laws of mechanics and gravitation reign universally. In fact, truly cosmological studies are rare in the nineteenth century;* the most remarkable are those

* Aside from the theories of Kant and Laplace, all the hypotheses examined by Henri Poincaré in his famous *Leçons sur les hypothèses cosmogoniques*

that anticipate modern theories by revealing the difficulties encountered by any attempt at cosmological synthesis within the context of classical science.

And yet, in another sense, it is perfectly legitimate to say that classical cosmology exists, as the rise of modern cosmology has clearly revealed. It exists as a system of statements which support, more or less implicitly, the universal validity of the arguments of physics and astronomy; these statements have to do with space, time, nature, and the form of physical laws. For physicists and astronomers never doubted, before the twentieth century, that the universe possessed certain structural properties which were clearly defined and strictly consistent among themselves. In fact, the elements of this more or less implicit structure for two centuries served as the foundation for the whole impressive apparatus of physics and astronomy—and of that "formal" cosmology which we shall try to make explicit. So that the reader will bear in mind this ambiguous quality of the theory of the universe in the classical age of science, we shall speak instead of classical "pseudo-cosmology."

I. GEOMETRY AND THE REIGN OF EUCLID (Ia, b, c)

More than Greek cosmology, classical science is based primarily on the soundest of the acquisitions of Greek science, geometry. Aristotle had explicitly refused to regard geometry, the science of abstract objects, as being of the same nature as physics; as a consequence, he saw no difficulty, on the one hand, in agreeing with the geometers that one can hypothesize about figures as large as one chooses, and on the other hand, in stating that the world is finite.*

Moreover, the Greeks, who invented geometry, had not completely formed the concept of *space*, or more accurately, had not established a very precise relationship between the concept of space and the science of geometry. This relationship is, however,

---

have the two characteristics of dating from the late nineteenth or early twentieth century and of being limited in practice to the question of the formation of the solar system.

* *Physics*, III, 207B.

at the foundation of classical science. For Galileo, as for Descartes and for Newton, as for all the theoreticians of physics until the end of the nineteenth century, the axioms which Euclid had placed at the foundation of geometry (Ib), the theorems which he and his successors had deduced from them, and all those which could ultimately be deduced from them have, in some sense, the validity of physical laws. At any rate, they create the frame of reference within which all physical laws are to be inscribed, as well as the model according to which they are all to be conceived.

This postulate of the essential pertinence of geometry to an understanding of the physical world, which is related to the firm belief that the axioms of Euclid are known by "natural light" and belong to the same level of certitude as the truths of arithmetic—this postulate may be regarded as fundamental to the classical pseudo-cosmology, and constitutes one of its cornerstones. It remained untouched by the controversies of rival theories; Cartesians, Leibnitzians, and Newtonians all accepted it; it informed the gradual extension of the methods of mathematical physics in the nineteenth century. Even the demonstration by the geometers of the eighteenth and nineteenth centuries that Euclidean geometry is only hypothetically necessary did not, in over fifty years, make a dent in the confidence of physicists that this postulate holds true. Until the beginning of the twentieth century, or, let us say, for the sake of simplicity, until Poincaré, the position of physicists with regard to non-Euclidean geometry was analogous to the position of Aristotle toward pre-Euclidean geometry. An abstract science dealing with objects existing only in the realm of mind, the propositions of non-Euclidean geometry said nothing, according to them, about real space, and did not express the geometrical properties of bodies: the world of physics was Euclidean.

CONSEQUENCES

The world is infinite. This cosmological consequence of the interpretation of Euclidean geometry as the fundamental science of nature is immediately evident and, from the outset, disastrous to the ancient system of the world. At any point in space, it is permissible to draw three straight lines, each perpendicular to the other two, and to extend each one in both directions. The space of the geometer is infinite in all three dimensions; therefore, the world is, too. The result is that the celestial vault is only an illusion, a cosmic horizon which reproduces the mirage of the terrestrial horizon.

Galileo, for whom geometry was more a method than a metaphysics, ultimately hesitated before the dogmatic affirmation of the infinity of the world; but his astronomical observations had offered striking proofs of the enormous depth of the universe in this third dimension, so difficult for astronomers to investigate. Notably, the resolution of the Milky Way into separate stars proved that it is an enormous accumulation of stars so numerous and so remote that the eye cannot distinguish them.*

Descartes, fortified by the observations of Galileo and convinced for metaphysical reasons that geometrical knowledge reveals the essential properties of bodies, did not hesitate to state dogmatically that the universe is infinite. A few years after Descartes' death, Pascal was no longer able to contain his anguish before the idea of the infinity of the cosmos by returning to the more reassuring image of the perfect sphere centered on the earth.

The world has no top, bottom, center, or circumference. The figures of geometry are unaffected by changes in position or rotations. To slide a cube along the plane of one of its faces or to turn it upside down of course changes its appearance for the unmoving spectator, but does not alter any of its properties. In the space of geometers, therefore, there is neither center, nor circumference, nor top, nor bottom, nor right, nor left, except by convention. If one calls a certain direction the "top," then the opposite direction will be the "bottom"; but no direction is in and of itself the top, and any point may legitimately be taken as the center. Consequently, there is no real center to the universe,

* *Sidereus Nuncius*. Italian edition and translation by Maria Timpanaro Cardini, page 42.

there is no absolute orientation in the world. The vertical direction is not opposed to the horizontal, or is only opposed to it relative to the presence of the earth and relative to the position which gravity assigns to mountains, to plants, to our bodies, to what we carry and build with our hands.

Here too the observations of Galileo confirm the reality of that decentering of the universe which Copernicus, after Aristarchus, had imagined, and thus encourage the hypothesis that identifies real space with a mathematical Euclidean space. For Galileo observes with care, and without possible error, four small planets revolving around the large planet Jupiter: an argument of weight in favor of Copernicus, as he remarks himself, in a very significant passage which reveals the difficulty his contemporaries had in renouncing the traditional concept of cosmic motions.* Many people, Galileo writes, would be willing to accept the Copernican system, but they cannot understand how the moon could be the only body that revolves around the earth, and how the earth and the moon could revolve together around the sun. In other words, the lunar rotation, so obviously real, seems to restore to the earth its central position and annoyingly confuses the Copernican image of the sun as the center of the world. But Galileo *saw*, with his telescope, the satellites of Jupiter revolving around that body just as the moon revolves around the earth, and at the same time he saw Jupiter drawing its satellites with it in its rotation around the sun, just as the earth draws the moon with it. The evidence of the senses is now in favor of this multiplication of centers in the universe.† The earth and the sun may therefore be, as Jupiter evidently is, relative centers of particular rotation, even as they are drawn along themselves in other movements. Centration, rotation, and satellization are only local and relative circumstances in infinite space, which is itself unaffected by the movements of bodies.

Geometry, whose fundamental assumptions (infinite, noncentered space, interchangeable directions) so strongly contradicted cosmic appearances (the spherical nature of the world,

* *Op. cit.*, page 80.

† *Sensus nobis offert vagantes stellas . . .* , *op. cit.*, p. 80.

absolute orientation, etc.), was nevertheless itself of experimental, even technical origin. Indeed, this may be the ultimate reason for the triumph of geometry in a civilization which, from the seventeenth century on, was engaged in a great technological adventure. Surveying and the demarcation of land—so necessary to the nonnomadic peoples of Egypt, where the flood waters of the Nile regularly invaded the land and erased the boundaries of the fields—and, even more, that great architecture which for thousands of years, from Persia to Spain, from the Sahara to Hadrian's Wall, gave the ancient world the most impressive feature of its physiognomy, had combined to produce geometry as a kind of subtle and marvelous efflorescence of the labor of men. Out of all these tasks of surveyor, stonecutter, mason, had emerged the supreme art that governs them all, the art of measurement, which geometry carries to the pinnacle of theoretical abstraction.*

Geometry, as its name tells us, is a metrical science, a science of measurement (Ia). And one of the most important aspects of the geometrization of physics, of the promotion of geometry to the rank of a fundamental science in the understanding of nature, is that physics itself becomes a metrical science. To know the universe is first of all to take measurements and to combine these measurements by means of mathematical relationships which are appropriate to express them. Classical science postulates that the world in which a physics of measurement is rooted is so constructed that this metrical enterprise is within the reach of rational knowledge. Of course, Greek astronomers had already applied this procedure to the study of celestial phenomena, but this early mathematization of nature remained partial: it did not extend to the earth. Moreover, it remained descriptive and did not concern itself with the causality of motion. In classical science, on the other hand, the mathematization is, by its own right, total. It is extended to the whole universe, and it involves the explanation of phenomena as well as their description. Relations of causality were themselves to be made mathematical in that form referred to as a law, to which we shall return.

But that assumption of the classical system which grants to

* Cf. E. Husserl, *Die Krisis der europäischen Wissenschaften*, II, § 9*a*.

physics *a priori* the possibility of its operations of measurement and of their consistency is the assumption that space, and, therefore, the geometrical properties of bodies, are in harmony with the Euclidean axioms (Ib). From this follows a large number of consequences, which it would take too long to enumerate and discuss; let us simply note the most important from the point of view of cosmology:

▪ Space is a universal container whose metrical properties are absolutely invariable and totally independent of time, place, or material content. One can, therefore, choose a unit of length which will maintain itself in time and space, unaffected by changes in position, or by the passing of time, or even by all the transformations of the universe (Ib3). This unit can be made tangible in the form of a solid which will be protected from any deforming physical or chemical influences that might be exerted on it.

Following the international convention in 1875, the installation at the Pavillon de Breteuil in Sèvres of the famous meter made of platino-iridium marks the apogee of the classical philosophy of nature. Preserved from all cosmic contingencies (within the limited scope of the powers of man), the Meter of Sèvres was the materialization of a geometrical reality absolutely protected, by postulate, from all the contingencies of time and place. According to the conception of nature underlying classical science, a cosmic cataclysm of minor importance could, of course, reduce the Meter of Sèvres to dust, just as time had reduced its creators to dust, but no cataclysm and no passage of time would ever destroy the geometric meter of which the Meter of Sèvres is the material embodiment.

▪ In the Euclidean system, the shape of objects is independent of their size; the dimensions of objects of the same shape can be deduced one from the other, provided one knows the ratio of the dimensions of two or more of their corresponding elements (Ib4). In this manner a very large triangle can be measured by starting with measurements made on a small triangle. It was in this way, according to tradition, that Thales, by making measurements on the coast, calculated the distances of ships at sea. This is extremely important, not only for geodesy but for astronomy and, through astronomy, for all of physics, for so long as we have a base whose size we know it enables us, by simply measuring the angles, to calculate the distances of celestial objects, to measure the stellar system, and ultimately, the system of the galaxies in three dimensions. We shall see that one of the great difficulties of modern cosmology is that, in the case of the very great distances, it no longer has this necessary base at its disposal.

To say, as Kepler did, that Mars describes an ellipse of which the sun occupies one of the foci would be perfectly arbitrary and would even, strictly speaking, be meaningless, if this Euclidean reduction of similar figures to one another were not possible.*

The science of space is not the only foundation of the classical system; this system also rests on postulates of equal importance relative to time (II). But these postulates are more difficult to formulate, for there is no such thing as a "science of time" comparable to geometry. The word "chronometry" refers more to a technique than to a mathematical science, as the word "geometry" does. More accurately, as Aristotle had seen,† the science of time is the science of number. But the founder of this science is neither Aristotle nor Euclid, but Newton, who formulated, in terms as celebrated as they are concise, the postulate which underlies the position of classical science concerning the representation of time: "Time is a number for movement according to anterior and posterior. Time is not mathematics but is expressible by a number."‡

This means that the clock measures something that is independent of it, just as the meter measures something that is independent of it, something that will continue to exist if the clock stops working or even if it is reduced to dust. It also means that, just as it was possible to install at Sèvres a meter that was "perfect" in terms of human capacities, one can also try to construct a perfect clock that will keep the "same" time indefinitely; something, in short, that is everywhere and always identical to itself, like the unit of length.

* The first scientific measurements of antiquity—the measurement of the distance from the earth to the moon by Aristarchus, and the measurement of the radius of the earth by Eratosthenes—were already based explicitly on this postulate.

† *Physics*, IV, 219B.

‡ *Principia*, Book I, "Definitions."

From the point of view of the classical system, the primacy of the interval of time as an objective reality, which concrete measurement is seeking to attain, is just as fundamental as the postulate of the Euclidean properties of real space. For this concept of time assures the validity of the mathematical structure of physics. In fact, after Newton came the gradual establishment of the type of mathematical expressions that were to become characteristic of physics. The most fundamental laws of physics define certain quantities as functions, in the mathematical sense of the word, of a numerical variable which in fact represents time. Newton's postulate about absolute time signifies that this mathematical variable has, in some sense, an exact and unique referent in nature, no matter how imperfect the measurements we make of it.

The explicit rejection of this postulate by Einstein in 1905 was to be one of the first acts of the second cosmological revolution.

II. HOMOGENEOUS MATTER AND THE PHYSICS OF LAWS

Between what Aristotle and his successors called "physics" and what has been called by this name since the eighteenth century there is only a rather vague connection, precisely because of the almost negligible role played by geometry and chronometry in the first, and the extremely important role that they play in the second. And yet it is not meaningless to say that the classical notion of the universe rests much more clearly than did the ancient cosmology on a form of physics—that is, on a system of rational interpretation of the phenomena which we observe not only in the heavens, but in our immediate natural environment. Whereas in the first "physics" the stars are not made of the same substance as terrestrial things, and celestial phenomena are not understood in the same way as terrestrial phenomena, the second physics—classical physics—establishes as a postulate that matter is everywhere the same and that all the phenomena of the universe obey the same laws.

This postulate is almost unbelievably audacious in the extent to which it extrapolates experience. One is sometimes inclined to wonder at the frequency with which the last seventeenth-century supporters of the ancient physics invoked experience against the new ideas. But we find that while Galileo's observations led him to deny that there was any absolute difference in kind between celestial objects and terrestrial objects and, in particular, to refute the alleged incorruptibility of the stars, he was, on the other hand, very far from being able to present the slightest evidence that the whole universe is composed of the same matter.

As for Descartes, he thought that on this, as on many other questions, he could dispense with experience. He knew, by "natural light" and by divine right, that matter was not distinguishable from the homogeneous "extension" of geometry and that consequently it is everywhere and always the same, in spite of the diversity of its forms. What distinguishes Descartes from the rest is the degree of his intellectual arrogance; for the atomists of his day, his adversaries, and all those who came after him thought no differently from him on the question of the uniformity of matter, and they also had no proof. In fact, this postulate underlies all of classical physics and was accepted without the slightest suggestion of real evidence. This proves that we are dealing with a veritable exigency of reason, which took over the moment that rejection of the geocentric hypothesis gave rise to the risk that one might see the unity of the cosmos dissolve completely in a "plurality of worlds" and reason lose its control over the universe.

It can now be said that the Galilean wager has been won and that reason has been rewarded for its audacity. Since the last century, astronomy has been joined by a new science, astrophysics, which, by making use of the findings of experimental physics, has at its disposal some very effective means of gathering information about the matter of which extraterrestrial objects are composed.* And so far at least, all the pieces of information gathered agree on one point: there are in the universe no other

* It is the spectrum analysis of light and radio and X-ray astronomy that provide the totality of our information about objects located beyond the solar system. In the case of objects within the solar system, spectroscopic information is supplemented by information obtained by the examination of meteorites, the debris of celestial objects that have fallen to earth, and, more recently, data furnished by materials taken directly from the moon and by close observation of the nearer planets.

species of atoms naturally occurring than those of the ninety-two elements, from hydrogen to uranium, with their 325 isotopes, that are known on earth. On several occasions, by means of spectroscopy, elements have been identified in the sky that were unknown on earth (this is why helium, first observed in the solar spectrum, bears its name); but these have always turned out to be elements that were in fact present on earth, but had not been detected, or elements that were well known, but with previously unidentified properties.

For cosmology, moreover, the universality of laws is even more important than the homogeneity of matter. The first physics, properly speaking, had no laws, and did not express itself in terms of laws, not even in the domain where the scientific method and spirit already reigned, astronomy. There are no "laws" discovered by Eudoxus or Ptolemy, and Archimedes is no doubt the only physicist of antiquity of whom certain propositions deserve, in the eyes of moderns, the name of "laws" or "principles"; but as we have seen, Archimedes had little interest in cosmology.

Classical physics, on the contrary, presents itself as a system of laws.

What does this mean? As we have already briefly suggested, the organization of experience by means of laws is inseparable from the metrical and hence mathematical character of classical physics, in which a phenomenon is characterized by a set of parameters whose measurement enables one to describe and identify it. If direct measurement is not possible, it may be replaced by calculation. A law is therefore a functional relationship (functional in the mathematical sense of the word) among these dimensions, a relationship which is maintained among them and which one encounters in all phenomena of the same kind. This may be—and this is certainly the most important case for physics—a relationship which indicates how a certain quantity varies over a period of time. For example, according to the law of falling bodies, which was discovered by Galileo, the distance of a fall varies with the square of the time of the fall. But actually the most general laws of classical physics are based on a kind of limiting analysis of phenomena. In other words, the laws mathematically describe the relationships that exist among the physical dimensions when one considers these dimensions at intervals of time and space that are arbitrarily small. This procedure, invented by the great physicists and mathematicians of the seventeenth century, particularly Leibnitz and Newton, has turned out to be extraordinarily effective in spite of its abstract and apparently nonexperimental character, because in principle it describes phenomena to a degree of infinite fineness which is by definition inaccessible to experience.

To express these laws conveniently and to make them suitable to describe and calculate phenomena mathematically, it was necessary to invent an adequate mathematical theory, differential and integral calculus, by means of which an apparently confusing idea became practical. The idea is that of effects, instantaneous in time and at a point in space, which change from moment to moment and which are produced by causes that are themselves usually variable but which act continuously. These laws take the form of differential equations [for example, the fundamental law of dynamics (III b 1, 2)] or equations with partial differential coefficients (for example, Maxwell's equations of electromagnetism).

THE REIGN OF MECHANICS (III)

It was to mechanical phenomena that the new methods were first applied, with a success which by the middle of the eighteenth century was already so evident that it seemed to have determined forever the future course of physics, without room for hesitation or ambiguity. A definition that would indisputably distinguish a mechanical phenomenon from all others would by no means be easy to formulate. However, the motion of a body, small or large, which has a mass (that is, which resists an effort to move it or stop it and which retains its identity as it moves) is a mechanical phenomenon. Thus, the moon's motion is a mechanical phenomenon; the propagation of light is not. According to the nature of the forces that are causing this motion, we may be dealing with a purely mechanical phenomenon (if the motion is produced by gravity, a collision, the action of wind, or another movement) or with a phenomenon which is also of another nature (if, for example, the approach of a magnet causes an iron needle to spin). In certain cases, the same phenomenon may be regarded as mechanical on one scale and not mechanical on another. For example, the heating of a bar of iron by conduction is not a mechanical phenomenon; but this phenomenon may be described physically as the movements of atoms and is therefore, from a certain standpoint, mechanical.

To clarify these ideas, let us take the example of a phenomenon that is typically mechanical: a solid object moving through the air, such as a javelin being thrown by an athlete. This phenomenon is described and explained in terms of physics, by assigning numbers, first to the measurable elements characteristic of the solid—its mass, its length, its shape, the density of its various parts; then to the forces which have acted or are acting on it—the impulse provided by the athlete, the weight, the resistance of the air; then to the kinematic dimensions characteristic of the movement itself—the time, the rotation of the javelin around its center of gravity, the movement of this center in relation to the earth, and the variations of these movements (their speeds and their accelerations).

The laws of mechanics relating to the movements of solid bodies which are applicable to the case in question are simply relationships among the numbers that measure these quantities and the possible variations of these numbers,* relationships which are regularly and universally verified by all those solid bodies on which it has been possible to conduct experiments. Since the values of these numbers are assumed to be known at the time the javelin is thrown, the laws enable one to predict its movement. But, at the level of fundamental mechanics, the laws governing the movement of the solid body are themselves merely the consequences of more general laws concerning the instantaneous movement—in response to the action of its material medium—of elementary bodies or material particles which are imaginary, being the result of a limiting analysis comparable to the one we

* In classical mechanics, masses do not vary. In the case of solid bodies, other numerical attributes also remain invariable.

have mentioned (IIIb1). Starting with these material particles and guided by experiment, mechanics constructs various systems (IIIb4), flexible fibers or taut cords, membranes, solids, liquids, gases. By combining the fundamental laws with the proper modes of connection between elements of these systems, mechanics can mathematically construct special laws governing these "models" (as they are generally called now), which controlled experiment will later verify and may possibly refine. Sometimes the elementary particles are separate and free, sometimes they are bound, while still remaining separate and distinct, sometimes they merge in some way to form a continuous medium, in which case the "material point" is replaced by the infinitesimal element of line, surface, or material volume, which is just as imaginary as the first. Indeed, mechanics is not overly concerned with the physical accuracy of these models, none of which represents exactly, by all appearances, the natural systems; in fact, several of them may serve to describe the same systems at different orders of magnitude.

It was consistent with the spirit of classical science, which sought the unity of the universe in the universality of its laws, to try to reduce these fundamental laws to the smallest possible number. Mechanics, together with optics, whose special role we shall examine further on, was the first of the physical theories worthy of that name. The great logical consistency of its principles, the fertility of its mathematical exegesis in finding equations for phenomena of increasing diversity and complexity, and above all, perhaps, its striking successes in the prediction of astronomical phenomena, that is, on the very terrain where the first cosmology had solidly established itself—all this contributed gradually to confirming the theory that all the phenomena of nature, once suitably analyzed, would eventually fall under the jurisdiction of mechanics.

Moreover, the success of mathematical mechanics seemed to confirm the authenticity of a certain basic model of nature, half-conceptual, half-intuitive, which had gradually replaced the qualitative and organic image that ancient science had formed of the terrestrial world. This new model was mechanistic, and Descartes had summed it up in a famous remark: "Everything is done by design and motion." It was by referring more or less to this formula, and to the experiments and images associated with it, that mathematical mechanics had gradually constructed its fundamental notions and elaborated its principles. The fertility of the mechanistic idea results partly from the fact that it is not necessarily embodied in a single model but allows for several. The simplest of these, which was invented in the early stages of Greek science, is the "atomic" model. Matter is imagined to be composed of identical and immutable atoms which are, however, able to move in all directions, and all the diversity of phenomena proceeds from the diversity of the combinations and movements of these atoms. It was from this model that mathematical mechanics borrowed its basic concept, that of the material point, an atom whose mass is assumed to be finite and invariable, but whose spatial dimensions may be regarded as arbitrarily small, so that this atom can be associated at any moment with a point in space. Another and much more complex model is the fluid, for which familiar images are provided by water and air. In this case matter is identified with a continuous and homogenous fluid in which currents and eddies, the buildup and release of pressure, trace configurations that are infinitely varied, stable or otherwise. This is the model of Descartes, but mathematical mechanics had more difficulty mastering the fluid than the atom. Pascal, Bernoulli, Euler, and d'Alembert were the chief contributors to this theory.

Thus, mechanical models guided mathematicians in the founding of mechanics, and then exercised their ingenuity in its development. In turn, the striking successes of this new science contributed in a decisive way to winning acceptance for mechanism, a hypothesis according to which, in the last analysis and after detours of varying length, the whole of nature is to be conceived and understood in terms of one or more mechanical models. Classical physics and pseudo-cosmology were, therefore, essentially mechanistic. Celestial mechanics was the principal contributor to this success.

Edmund Halley, an English astronomer who was closely associated with the genesis and publication of Newton's masterwork, and was a friend and admirer of its author, ventured to predict, in accordance with Newton's theory, the return of a particularly brilliant comet which had appeared in 1682 and which he had identified as that sighted by others in the past (Va3).

Several dozen years later, Alexis Claude Clairaut again took up the somewhat approximate calculations begun by Halley. Clairaut had at his disposal more powerful mathematical tools which enabled him to calculate with much greater precision the disturbing influence of the large planets on the comet. He predicted the return of the comet for the middle of April, 1759, with a margin for error of thirty days either way; the comet appeared on schedule and passed through its perihelion (the point on its orbit that is closest to the sun) on March 12, 1759. In the meantime, Halley had died.* The scientific importance of the event did not escape those living at the time; a half-century later, in a famous passage, Laplace, in the style of a typical "philosophe" of the age of enlightenment, hailed this shining victory of reason over the superstitions to which comets had given rise since the dawn of time. For Laplace, as for many of his contemporaries, mechanics, to which his senior the Piedmontese Joseph Louis Lagrange had given an almost perfect mathematical form and to which he himself had contributed some important theorems, was the fundamental science of civilization, the key to the universe, the liberating science.

As a consequence, mechanistic philosophy found itself the beneficiary of the irreplaceable imprimatur of mathematical mechanics. Henceforth the authority of Lagrange and Laplace superseded the authority, more metaphysical and hence more doubtful in the eyes of physicists, of Descartes.

Until the end of the nineteenth century, in spite of, or perhaps because of, an impassioned romantic reaction against this philosophy of nature, the thinking of physicists, as soon as it went outside the boundaries of specialized experience, tended to place itself under the jurisdiction of mechanics; if there were a system governing the world, it had to be mechanical. Numerous

* Halley's comet returns approximately every 77 years, which leaves very little possibility for the same astronomer to observe two appearances.

attempts were made to force into this framework, by means of hypotheses that were sometimes ingenious and fertile, but often unfounded and sterile, all the phenomena of the physical world—something that the whole world now agrees is impossible. The fact is that it was difficult for physics to retain the advantages of mathematical mechanics without also being subjected to the limitations imposed by all mechanistic models. For us in the twentieth century it has become evident that in fact we can retain those advantages. It was much less so for the scientists of the last century, who were accustomed to thinking, in the wake of eighteenth-century rationalism, that the mathematization of physical science was inseparable from mechanistic philosophy. The creation of a mathematical physics that was completely liberated from the domination of mechanics was at once the last idea of classical science and the first act of modern science.

OPTICS, THE HARMONY OF VISION AND SPACE (IV)

Optics, the science of the properties of light, was associated just as early and just as lastingly as mechanics with the classical philosophy of nature. The fact is that up to a certain degree of precision in the analysis of the phenomena of light, optics is above all an illustration of geometry, and it provides the key to all observation that extends beyond the field of natural vision. As physicists see it now, the theory of light is an edifice with two stories which correspond to two levels on the analysis of reality. On the upper level, the one most easily accessible to experience, there is geometric optics (IVb); the idealization of the light ray, a pencil of light assumed to be infinitely fine, enables geometry to deal with problems relating to the propagation of light in transparent or reflective media, through the discovery, by experiment, of certain fundamental laws. These include the law of reflection, the law of refraction, and Fermat's principle, according to which light, which travels at different speeds depending on the medium, always follows the shortest path in terms of time. The deeper level, the level of physical optics (IVb), which is much harder to arrive at with any success by experiment, tries to describe all of the elementary phenomena involved in the emission, absorption, and propagation of light. Seen from this fundamental level, geometric optics represents an approximation which is valid only for certain phenomena and at a certain order of magnitude. But this approximation has the great advantage of being very largely independent of any hypotheses that may be made about the physical nature of the elementary phenomenon. And it is adequate for a good percentage of the problems that have to do with vision and its improvement by means of optical instruments.

The historical development of optics reflects this division. The fundamental laws we have referred to were known by the middle of the seventeenth century, and geometric optics was able to develop extensively without being seriously hindered by the disagreement that persisted until the beginning of the nineteenth century between those who thought light should be conceived as a stream of particles and the more perspicacious advocates of the wave theory.

Geometric optics helped to give classical astronomy its foundations in positive science. It contributed in two ways to the classical vision of the world. In the first place it was the instrumental science of this vision (IVb, c). The extraordinary Galilean opening-up of the sky would have been impossible without the telescope, and until the twentieth century, all the great advances of astronomy were related to advances made in optical instrumentation. It was Herschel's giant telescope that made possible our modern expanded view of the world of the stars, and it was another giant telescope which, as we shall see further on, initiated the cosmological revolution of the twentieth century.

But astronomy is not the only area involved; from the beginning of the classical age, geometric optics had been serving microscopy as well as telescopy, and it was over the whole vastness of the earth that it was furnishing proof of its effectiveness. From then on, this effectiveness ceased to be merely the result of skill and chance: it verified the geometrical postulate of an essential similarity between the infinitely large and the infinitely small; it proved that the Pythagorean theorem unites the two infinities. This explains why the contribution of optics to the classical philosophy of nature could not be limited to its instrumental role, for its success certainly contributed to assuring the hold that geometry, particularly Euclidean geometry, had over this philosophy.

The instruments of optics inspired an unqualified confidence in the supplementary information they provide about the microscopic world as well as the astronomical world.* This promotion of optics to the rank of a key science for observation refers us back to something more essentially theoretical, something that is already familiar to us: the fundamental suitability of Euclidean geometry to the description of the physical universe. For this suitability seems to be confirmed and supported by the fact that mechanics and optics employ the same structural elements and in the same order of increasing complexity. In mechanics, the straight line is the path of inertia of the elementary body; in optics, it is the light ray in a vacuum or a homogenous medium. In geometry, the conics (the circle, the ellipse, the parabola, the hyperbola) represent the second degree of complexity after the straight line; they were already well known to Greek geometers. But it is the conics that solve the simplest problem of celestial mechanics, the path of an isolated planet attracted by a sun according to the law of universal gravitation (Kepler's first law). Finally, if one rotates the conics on their axes, one gets surfaces which solve, at first approximation, the fundamental problem of dioptrics, a problem which fascinated the young Descartes and which he solved: how to use refraction to concentrate into a focus (in the seventeenth century, it was called a "burning point") the light rays emitted by a pinpoint source.

The very form of certain laws reinforces the evidence for this triple agreement among geometry, mechanics, and optics. Let O be a point in space, regarded as a solid center and a source of light (a star seen at a distance is a point of this kind). If one moves away from O by a certain distance R, the force of gravitation that O exerts diminishes at a rate of $1/R^2$ (mechanics), that is, in inverse proportion to the area of the sphere whose

* This confidence was not acquired in a day. The contemporaries of Galileo greeted the revelations of his telescope with some mistrust; just as doctors, at least until the middle of the nineteenth century, showed a great deal of skepticism about microscopic images.

center is O and whose radius is R (geometry) and in the same proportion (if one is in an empty space) by which the luminosity one receives from it decreases (optics).

ASTRONOMY: AN INCOMPLETE PICTURE OF THE UNIVERSE (V)

In the classical pseudo-cosmology, as in the ancient cosmology and in the modern, it is, of course, astronomy that ultimately sets the tone. But if classical cosmology is only a pseudo-cosmology, if it does not culminate, as ancient cosmology did or as modern cosmology does, in a true system of the world, this is essentially for two reasons which have something to do with astronomy and its place among the sciences. In the first place, astronomy was demoted from the ruling science it once was to an applied science. From now on it is the theorems of geometry, the laws of physics, and particularly, as we have just seen, the laws of mechanics and optics that are indispensable to astronomy in the conducting of its observations and the prediction of their results. Inversely, of course, the observations of astronomy provide the laws of physics with some valuable verifications, but it is not these observations that define the structure of the world, it is not they that constitute the final authority in the search for rationality in nature. Advances in physics tend to shift this final authority, experimentally, to the scale of the laboratory and, theoretically, to the level of the event point.

Second, the image of the world that astronomy can provide, within the limits of the means available to it at the end of the nineteenth century, is an image that acknowledges itself to be incomplete, unfinished, without a well-defined structure, indicating uncertainties beyond its scope. Without attempting to retrace, even in its broad lines, the fascinating and glorious history of astronomy between the beginning of the eighteenth century and the beginning of the twentieth century, it is nevertheless germane to our purpose to show how in spite of and, in a sense, because of, its enormous advances, classical astronomy at the threshold of the twentieth century had arrived at a vision of the world whose most striking feature is its incompleteness.

The observations of Galileo had definitively confirmed the

conclusions that Archimedes had correctly drawn from the heliocentric hypothesis of Aristarchus. Between the dimensions of the solar system and the dimensions of the stellar world (the world of the "fixed" stars) there exists an enormous difference in order of magnitude. So enormous, in fact, that after Galileo it took astronomy over two centuries before it was able to estimate approximately the distance of a star from the earth. However, during this period, or around the middle of the nineteenth century,* exploration of the stellar world had already begun, by new methods which were later to come into general use by all of the sciences: the methods of statistical investigation.

The founder of stellar astronomy is William Herschel. Confronted by the incredible profusion of heavenly bodies revealed by the telescope, Herschel managed to orient himself with the aid of the only two elements at his disposal: the grouping of the stellar images on the celestial sphere, and the differences in apparent luminosity (in "magnitude") among these images, differences which can be interpreted, on the average and up to a certain point, as resulting from differences of distance.

For a century the work begun by Herschel was carried on and perfected. By 1920 his successors had done such a good job of completing it (the last of the great classical scientists in this sense was the American Harlow Shapley), that just as the second cosmological revolution was about to take place, the essential incompleteness of the classical universe had become very apparent.

This incompleteness consists not in the fact that there remained an infinite number of things to be learned (in this sense, *every* system of the world has been, is, and, in all probability, always will be, incomplete), but rather in the fact that science possessed no certainty about the properties of the universe in terms of the *overall* distribution, in space and time, of matter and energy.

* The first stellar distances measured were the distances of Star 61 of Cygnus by Bessel and of Vega by Struve, around 1840.

If one remembers that the first accurate measurement, by order of magnitude, in the solar system was Aristarchus' measurement of the distance of the moon, one realizes that it took *twenty-two centuries* for the ratio between the orders of magnitude of the solar system and the stellar system to be determined.

What was in effect the physiognomy of the universe at which classical astronomy had arrived at the moment of its greatest advancement, around 1920?—an enormous mass of stars, whose dimensions and shape had been rather well determined, immersed in a medium which was empty or whose content was totally unknown. Of vast but ultimately finite dimensions, in shape rather regular, lenticular, with certain known elements of symmetry, but not perfectly regular—such was the appearance of our galaxy, the only system actually known and observed with any certainty at that date.

To be sure, the question of whether all the objects whose images are visible on the celestial sphere, particularly the images of nebulae, do or do not belong to our galaxy had been discussed for a century and a half, but it had not yet been completely resolved, and as a matter of fact, when this was finally done, the classical era came to an end. But so long as no certainty had been achieved on this point, so long as the galaxy could continue to be identified with the observed universe, classical cosmology could only recognize itself as incomplete. Indeed, this fact revealed a marked incongruity between the *container*, the Euclidean space—infinite, homogenous, isotropic—and the *content*, the galaxy, this material system—finite, "local," and only seemingly regular.

This situation was compounded, again in 1920, by an abysmal ignorance regarding the temporal dimension of astronomical systems. After the precise mathematical formulation of the laws of "energetics," no one could any longer doubt that the sun and the stars, as enormous consumers and transformers of energy, undergo an irreversible evolution and participate in a real history. The galaxy in turn, without any doubt, was a system in motion, which was in its turn composed of subsystems, themselves in motion; but before science could explain the evolution of the stars, the history of the sun, the history of the solar system, much less the history of the galaxy, certain essential elements were still missing.

One problem, however, was a partial exception to this helplessness of classical science in the face of the history of the cosmos: the problem of the formation of the planetary system around the sun. At the end of the eighteenth century Laplace, rejecting the doctrine to which Newton himself subscribed that the present arrangement of the solar system represents a special intention of the creator, had proposed an explanation for this arrangement in terms of the laws of mechanics. This explanation appears at the end of his "system of the world," which is, in fact, a theory of the solar system. The habit has persisted to the present time of using the word "cosmogony" to refer to any theory about the formation of the solar system—a misuse of language which is revealing, both of the limitations and of the nostalgia of classical science with regard to the evolution of the world. "Cosmogony" means the genesis of the *universe;* to apply the term to a theory of the solar system betrays an undeniable megalomania.*

But it was already a great triumph to be able to reconstruct the probable history of that gigantic formation that had served as a natural laboratory for mechanics. The general theory of Laplace, according to the verdict of our contemporaries, is probably accurate. Out of a large nebula, which simultaneously turns on itself and contracts under the effect of gravitation, are formed by successive fragmentations the sun at the center, and the planets in concentric rings.

But for Laplace and for all classicists until the twentieth century the problem was ultimately insoluble because it is beyond the means of mechanics, and because its solution, which is still far from being complete, requires that there be brought into play both electrical and magnetic forces, not to mention nuclear phenomena without which the radiance of the sun would be incomprehensible. This example shows that the advances in knowledge had themselves helped to accentuate the incompleteness of classical astronomy. The fact that it was possible even to approach the problem of the genesis of the solar system with the methods of classical science served to point up by contrast the inability of this science to raise the other questions of cosmology. Among these are the formation of the stars and of the galaxy, to say nothing of the atoms themselves, about which little was

* The same misuse of language occurs when we give the name "astronauts," let alone "cosmonauts," to spacemen who never leave the outlying regions of the earth.

known and which, as we shall see, have now, in our century, been thrown into the crucible of cosmogony.

Thus classical science could only proceed *a priori* and with few guarantees to a kind of formal summing up of the universe, by accepting Newton's postulates on time and space, by assuming Euclidean geometry to be universally valid, and by postulating the unity of matter and the universality of laws. Within this sterile frame of reference, science was at a loss to explain in concrete terms the whole of a universe. This is why between 1543 and 1920 we can only talk about a pseudo-cosmology, in comparison with the "true" cosmologies—the systems of Eudoxus, Aristotle, and Ptolemy on the one hand, and the contemporary systems on the other hand. We have explained what foundations underlay the rational consistency of the ancient cosmology. As for the new cosmology, we shall see that it is distinguished from the classical one both because it refuses to accept, *a priori*, that the metrical Euclidean-Newtonian form is the appropriate one for the universe, and because it interprets a new, observed concept of the universe which makes legitimate, or at least makes plausible, the always hazardous operation of generalizing.

CONCLUSION: FORMAL COSMOLOGY, THE "SEDIMENT" OF CLASSICAL THEOLOGY

Thus the classical science of the universe maintained itself for three centuries without any solid foundation; for we must now return to that astonishing act of audacity by which the great founders of classical science began by assigning structural properties to a universe about which the entire body of their own principles and their own observations proved that they knew nothing. A few remarks by Newton show that he had assessed the profundity of the ignorance into which the shattering of the ancient cosmos had plunged the philosophy of nature; and Descartes reasonably refrained from prying into the intentions of the creator. But this did not prevent the one from affirming *a priori* and dogmatically the infinity of the world, or the other from talking about time and space as "absolute, true, and mathematical." What reasons drove both of these men and, for that matter, almost all of their contemporaries to a dogmatism so intemperate, so manifestly contrary to the rules of prudence which each in his own way recommended to all who wish to philosophize? The answer is "God."

Not that the masters of the "great rationalism" were in any better agreement among themselves about God and his relation to nature than they were about the fundamental properties of any physical world; Descartes refused to consider final causes, that is, the supposed elements of the plan that God may have had in mind in creating the world. On the other hand, these final causes are an essential part of the system of Leibnitz, and Newton does not deny himself the right to appeal to them. Spinoza rules out any remaining personal element in the God of Descartes, while Newton reproaches this same Descartes for depriving the Lord of that which makes him a person. But all are in agreement on one point, that the world would be unintelligible if we did not regard it as the work of, or perhaps a part of, a God endowed with an infinite existence and a sovereign reason. A certain affinity, in some sense congenital, between our limited human understanding and the infinite divine understanding permits the philosopher to affirm *a priori* the most fundamental laws of nature.

". . . I have observed certain laws which God has established so firmly in nature and which he has imprinted so steadfastly in our souls, that after reflecting on them long enough, we can no longer doubt that they are precisely observed in everything that happens in the world";* so writes Descartes. Newton, although he may not claim to have learned the laws of nature directly from God, nevertheless thinks that the arrangement of the world cannot be conceived without his design: "This admirable arrangement," he writes, "of the sun, the planets, and the comets can only be the handiwork of an all-powerful and intelligent being."†

The belief of classical science in the solidity of the rational framework of the world is, without any doubt, of theological origin. It is therefore most remarkable that this belief survived

* *Discourse on Method,* Part Five.
† Newton, *Principia,* Book III.

the gradual erosion of its foundation. For whatever opinion one may profess about the history of religious belief between the eighteenth and the twentieth centuries, one will certainly acknowledge that it tended to become more and more separated from scientific thought without, of course, being always opposed to it. It was less and less to the world, and more and more to the soul, that believers turned to find signs of God.

No doubt one could find somewhere in the literature of the nineteenth and even of the twentieth century statements comparable to the ones we have deliberately selected from the works of Descartes and Newton, but they would certainly not be from the pens of writers who played a role in modern science comparable to the one Descartes and Newton played in classical science. But the ebbing of theological thought had left science with a solid sediment. Euclidean space, Newtonian time, and the universal laws of nature remained the object of a dogmatism which seems as naive as the theological dogmatism of its founders. Moreover, the new rationalism, instead of engaging in the traditional self-destruction of philosophy, had long been forearmed against the consequences of this retreat of God from the world. Indeed, Kant was so anxious not to compromise the God of Faith in the eyes of the world that he transformed the fundamental elements of Euclidean geometry and Newtonian physics into *a priori* forms of knowledge. It was an excellent way to disengage physics from theology and at the same time to protect it from the risk of troublesome experimental encounters, yet without apparently drawing on any source other than experience.

But the wind of reason was blowing so hard at the end of the eighteenth century, and the triumph of mechanics was so dazzling, that the great masters of science did not always burden themselves with the scruples of philosophers. They did not hesitate to combine a dogmatic affirmation of the principles of classical science with indifference toward its theological foundations, or indeed with a rejection of these foundations.

Pierre-Simon Laplace, the son of a Norman peasant, was rescued from obscurity by his village priest and placed at twenty by Jean d'Alembert, who had detected his genius, in the post of professor in a royal military school. Having become citizen

Laplace, he was carried to the pinnacle of Republican honors by the Convention, then to the peak of imperial dignity by Napoleon, and the Marquis de Laplace was honored in his titles and prerogatives by the royal restoration. Visibly indifferent to the political reversals that accompanied the advance of Enlightenment, he was likewise indifferent to the dangers posed to the foundations of physics by a skepticism which, rejecting once and for all the theological arguments, might have reminded physicists that there is no experimental proof that the Pythagorean theorem or the law of inertia is valid in the vicinity of Vega.

The pseudo-cosmology of the classical age thus acquires in the nineteenth century the look of a formal cosmology made up of abstract pronouncements laid down dogmatically and *a priori* as universal, by a science which is less and less ignorant of its own ignorance about the concrete form of the world or about the final components of the structure of the physical world.

Let us recapitulate its elements. Space and time play the role of universal, immutable, and absolute containers, and the properties of space are Euclidean; the laws of mechanics and the law of universal gravitation were correctly formulated by Newton, and are valid for any time and any place; all the phenomena of nature must be reducible to these laws, for they must in the final analysis behave in accordance with a small number of elementary mechanical phenomena—the movement of a particle whose mass is constant and which is regarded as a point, the movement of a fluid or the vibrations of a homogenous flux that is perfectly elastic, etc.

As for knowing how the universe as a whole is constituted, or what its overall configuration is, these are questions which are probably insoluble and which had best not be raised. The only thing that astronomy shows us unquestionably, beyond the frontiers of the solar system, is a very large number of stars grouped in a mass with a distinctive shape—the galaxy.

The contrast between the confidence bestowed somewhat imprudently on the principles of formal cosmology and the ignorance that was recognized to exist concerning the distribution of bodies in space and their evolution in time explains why classical science did not try to reconstruct a complete and coherent image of the universe so long as the gaps and inadequacies of this formal cosmology had not been brought to light. And until the actual reconstruction of a real cosmology in the twentieth century, little attention was paid to the fact that if one followed the principles of formal cosmology, it was in reality *impossible* to form a complete, coherent image of the universe as a whole.

If, therefore, the science of the nineteenth century in a sense prepared the way for the rebirth of cosmology, it was primarily in a negative and, in any case, inconspicuous way. This preparation took place in those rare cases when certain postulates of the formal cosmology were criticized and called into question; and also in those cases, still rarer, when the incompatibility between the formal cosmology and a real cosmology consistent with its principles was perceived.

Doubts about the fact that physical space was necessarily Euclidean appeared as soon as mathematicians became convinced that the abandonment of the postulate concerning parallel lines introduced no contradiction into geometry, and that consequently non-Euclidean geometries were logically possible.

As early as the beginning of the nineteenth century, Karl Friedrich Gauss stated the problem explicitly and tried to solve it by observation. The geodesic measurements that he had made reassured him: with the precision that was then available, the results confirmed the Euclidean theorems. This is not surprising considering that, if physical space is not Euclidean, any local deviations are sure to be of an order of magnitude very much below the precision of any measurement made on the terrestrial scale. Besides, Gauss had very few imitators. In 1854 Georg Riemann set forth in a very general way the problem of geometry in terms which, 120 years later, need very little correcting.* Riemann observed quite clearly that among the infinity of three-dimensional "quantities" that are mathematically conceivable, physical space is distinguished by certain properties, notably *metrical* properties, which can only be determined by experience. He showed clearly by what special hypotheses Euclid

* This problem is raised in Riemann's famous dissertation, *Über die Hypothesen welche der Geometrie zu Grunde liegen* (*On the Hypotheses Underlying Geometry*).

had managed to determine these properties, and how, by dismissing only one of Euclid's hypotheses—that bodies exist independently of place (in other words, that a change of position is not sufficient to change their form)—one can conceive of other metrical properties for space. Thus one of the principles of the formal cosmology, the attribution, *a priori*, of Euclidean properties to space, was shattered beyond repair. However—and this is a symbol of the indifference of nineteenth-century science to cosmology—Riemann himself explicitly states that his remarks on geometry are interesting only insofar as they concern the properties of space on the scale of the infinitely small, that area where relationships of causality are determined, whereas "questions about the immeasurably large are questions that are useless for the explanation of nature."* William Clifford, who spread Riemann's ideas in England, did not disagree, and Poincaré, who did more than anyone else to demonstrate and explain the perfect equivalence, in terms of logical validity, between traditional geometry and non-Euclidean geometry, nevertheless struggled, by means of a subtle argument about whose validity one may have serious reservations, to restore Euclid's geometry to its ancient privileged role. According to Poincaré, this geometry is not more *true* than the others, but is undeniably more *convenient*, and consequently it "has nothing to fear from new experiments."†

Thus, the very persons who called into question one of the most fundamental principles of the formal cosmology, that of a structure of space that was *a priori* Euclidean, also helped in various ways to exclude all cosmological significance from the results of their efforts.

The calling into doubt of Newton's postulates on time, however, was not anticipated at all. It happened in a way that was sudden and unexpected, but decisive, in Einstein's first dissertation on the theory of relativity.‡ Of all the principles of formal cosmology, this was surely the one that was most solid and most profoundly rooted in "common sense"—at least this is what

* *Op. cit.*, § 3.
† *La Science èt l'Hypothèse*, Chapter V.
‡ *Elektrodynamik bewegter Körper*, *Annalen der Physik*, Volume 17, 1905.

seems evident from the incomprehension with which, for a long time, the theory of special relativity was received.

On the other hand, doubts about the absolute value of Newton's principle of mechanics were as old as the theory itself. The principle of inertia, which had already been challenged in the eighteenth century by Berkeley, was challenged again in the nineteenth century by Ernst Mach, when mechanics was enjoying its great success. For the motion of a body due to inertia—which is not the effect of any force and which does not result from an interaction between this body and its environment—nevertheless produces "a force of inertia" which is revealed by everyday experience; according to the principles of mechanics, this is combined with other forces, just as the other forces are combined among themselves. For example, in an artificial satellite of the earth, the force of inertia purely and simply cancels out gravity (IIIb1, 2, 3). This situation is not very satisfactory from the standpoint of logic, first because inertia seems to be an absolute property of a body, whereas movement by inertia, rectilinear and uniform, is relative in the Newtonian system; second, because it assigns totally different sources to forces which in terms of geometry are combined as if they were strictly interchangeable.

As we have observed, the rapid advancement of the physical sciences in the nineteenth century was characterized, among other things, by the extension to all observable physical phenomena of the mathematical methods that had succeeded so well in mechanics. But at the same time that this extension was taking place, it became more and more apparent that the gap between the principle of the ultimate reduction of all physical phenomena to mechanics and the actual carrying out of this principle might be very great, and perhaps unbridgeable. This was the case in the study of electrical and magnetic phenomena, then in the study of heat, and in that more general science of which the science of heat is only one part, the science of energy. The forms of energy are multiple and it is by no means evident that all these forms can be interpreted in terms of invisible mechanical phenomena (although quantitatively they are all equivalent to mechanical energies).

And moreover, another fact had become quite obvious by

the middle of the century: although the energies called into play in physical transformations are always conserved in quantitative terms, with each transformation they nevertheless lose part of their availability for subsequent transformations. This fact is very difficult to grasp, given a purely mechanical representation of physical phenomena.

Finally, not only did classical science fail to pass from a formal cosmology to a real cosmology for want of astronomical information that went beyond the galaxy, and also for want of satisfactory hypotheses on the evolution of the great physical formations, but indeed it would have been prevented from so doing, for such a procedure would soon have led it to contradictions. In effect, if the galactic system is all alone in empty space, there is little possibility of its being stable, that is to say, little possibility that it will not eventually be dispersed. But since Newtonian time is infinite, this makes it totally inexplicable that this system exists *now*. If, on the contrary, one assumes (and this is precisely what modern cosmology seems to confirm) that Euclidean space is uniformly filled with matter as far as infinity, one comes up against genuine contradictions. The best known of these contradictions is the paradox of Olbers: if the stars, all of which are assumed to have the same average luminosity, are distributed uniformly around the earth in spheres of increasing radius, the number of stars on a sphere of radius R will be proportional to the area of the sphere, that is, proportional to R^2. But the radiation coming from each star will be diminished in proportion to $1/R^2$, with the result that the radiation received on the earth from each sphere will be constant and greater than zero. Since there is an infinity of spheres, the sky should be infinitely brilliant, even at night.

Like the geocentric system before it, the formal cosmology of the classical age was, therefore, more or less secretly undermined. But it appeared to be too well protected by the success of physical science to be in danger of collapse. And yet the rupture did occur, and in a rather unexpected way. It came not from astronomy, but from theoretical physics, and it struck the system at what was apparently one of its strongest points: the philosophy of time and space.

CLASSICAL SCIENCE AND PHYSICAL PHENOMENA

Scientific Account

I. GEOMETRY

Geometry, as an abstract science, describes the framework within which mechanical and physical events occur and assigns to this framework, by inductive methods, certain properties which are exploited by deductive methods. Classical physics postulates that all the properties of space that are translated into geometric language are absolutely independent of the phenomena that are taking place within that space.

The origin of geometry is entirely practical: measurement of the boundaries of fields, of the length of roads, of areas, etc. These are the goals that this science set itself, so if one wants to analyze its mechanism, one can begin with the following schematic considerations, which seem in the last analysis to have been the first preoccupations of geometers.

Geometry reduces the vague contours of things to idealized

objects, lines without thickness formed of points without dimensions, surfaces swept by lines, figures formed by lines on these surfaces, and finally volumes defined by surfaces.

Geometry measures two kinds of quantities, angles and lengths, which requires the possibility of moving lines by rotation to determine angles, and by translation of an elementary line called a rule to determine lengths.

A rule is a thin continuous line which is defined by two points A and B and to which one assigns a number, one, for example, which becomes by definition the length of the rule from A to B. It is assumed that this length is a constant number, no matter what the position or direction of the rule may be in space. It is further assumed that the rule is thin enough so that if one matches points A and B up with two points A' and B' of a given line, all points on the rule will coincide with the points on the line located between A' and B'. In other words, if this coincidence does not occur one divides the rule into as many parts as are necessary to make one of the parts obtained have the desired property. One might also assume that rule AB may be bent in such a way that all its points may be made to coincide with those of line $A'B'$, without this operation changing the quantity assigned to rule AB, referred to as the length of the rule. This is what happens, in fact, when one bends a measuring tape in order to measure the length of a curved line.

Now let us suppose that rule AB coincides with arc $A'B'$ of a line and that one moves the rule so that A now coincides with B'; point B will now coincide with a point C' of the line. If, for example, it was necessary under these conditions to move the rule four times between points A' and E' of the line, the length of the line between A' and E' is said to be four. If one cannot completely cover line $A'E'$ by a whole number of displacements of the rule, one uses a shorter rule, for example a rule whose length will be called ½ if it is necessary to move it twice in order to measure exactly the length of rule AB. This length will be called ⅓ if it is necessary to move the rule three times to measure the length of AB. Thus it is assumed that one can always find a rule small enough so that all its points may be placed in contact with as small an arc as is desired. Moreover, just as the square root of 2 is not a rational number, that is, cannot be expressed in the form of a fraction consisting of two whole numbers, so one will have occasion to consider lengths that will be expressed by irrational numbers.

a1. DISTANCES. Given two points in space O and O', let us imagine that we are drawing all the lines that join these points. Let L be one of these lines. If the length of line L from O to O' is 1, we say that the distance from O to O' along the line in question is 1. Thus we would say that the distance from Paris to Strasbourg along national highway number so-and-so is 520 kilometers (Fig. 16).

A line is a space of one dimension, in the sense that if one possesses a single parameter (for example, the distance on this

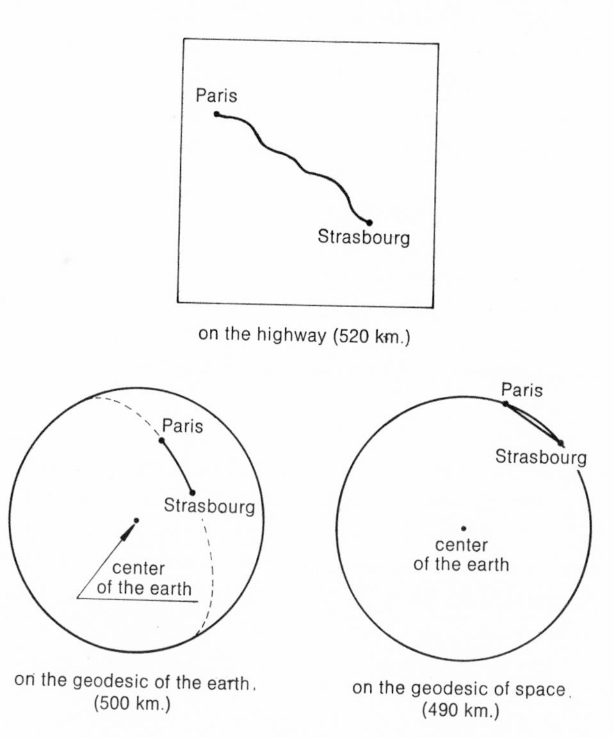

FIGURE 16

line of a point O' from a given point O), one is able to locate completely a point of the space. A space created by lines is a space of two dimensions if it takes two parameters to locate the position of a point on this space, as is the case with the surface of a sphere. A space of this kind is known in classic geometry as a surface. If two points O and O' are located on such a surface and if one considers all the lines on the surface that pass through O and O', the distance from O to O' on one of these lines will be shorter than it is on any other. A line possessing this property is called a geodesic of the surface. In the case of certain points on a surface (for example, two points O and O' diametrically opposed on a sphere), there can be an infinity of geodesics passing through these points. (All great circles on the sphere whose diameter is OO'.) On a geodesic of the spherical earth, the distance from Paris to Strasbourg is 500 kilometers.

If one needs three parameters to know the position of a point on a given space, this space is said to be a space of three dimensions. To describe the position of any point whatsoever in the universe of classical physics, it is sufficient to know three parameters, with the result that classical physics assumes the universe to be three-dimensional.

If now one joins O to O' by all possible lines in this space of three dimensions, and if one measures distance OO' along these lines, those lines for which the distance from O to O' is as short as possible will be known as geodesics of three-dimensional space.

b1. ESSENTIAL CONCEPTS OF EUCLIDEAN GEOMETRY. We say that the geometry of space is Euclidean if we accept as fundamental truths certain properties which will be called the axioms of geometry. We shall consider the geodesics of the spaces of two or three dimensions straight lines. We shall call a Euclidean space of two dimensions a plane. We shall use the following properties:

Through any two separate points on a Euclidean space, there passes one straight line, and only one.

Given a point O on a straight line and another point M, point M may be chosen in such a way that the distance OM on the straight line is as great as desired, point *M* being chosen on either side of point *O*. In other words, a straight line is infinite in both directions.

In a Euclidean space of two dimensions, there exist straight lines which have no common point. Such lines are said to be parallel.

In a Euclidean space of three dimensions, given straight line *D* and point *O*, a straight line passing through *O* and through a point *M* on *D* creates a plane, when *M* moves along *D*. This plane is unique. In other words, through a straight line and through a point there passes one plane and only one. One of the straight lines passing through *O* in the plane defined by *O* and *D′* is parallel to *D*, and it is the only straight line on this plane that has this property. Through a point there passes one line which is parallel to a given straight line, and only one. Two parallel straight lines define one plane and only one. Two straight lines that have a common point define one plane and only one.

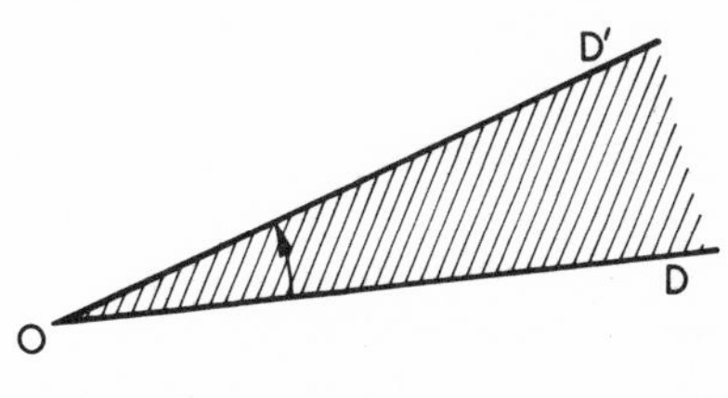

FIGURE 17

b2. ANGLES. In a plane, let us take a line segment *OD* and cause it to rotate around point *O* in a certain direction (Fig. 17). When segment *OD* has returned to its original position, we say that it has made one complete turn. Two segments *OD* and *OD′* which pass through *O* define one section of the plane called an angle, which is characterized by a quantity. This quantity measures the fraction of a full turn by which line *OD* must be rotated in order to bring it to *OD′*. An angle that corresponds to a quarter turn is called a right angle. Customarily a whole turn is divided into 360 parts which are called degrees, or into 2π parts ($\pi = 3.141592 \ldots$) which are called radians.

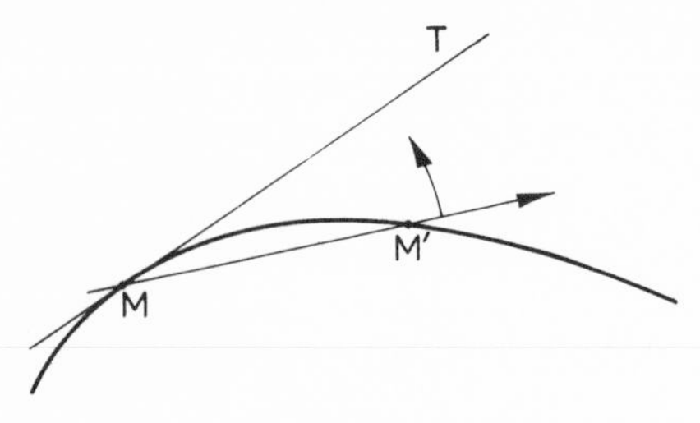

FIGURE 18

If we consider a curve and a point *M* along this curve, a straight line passing through *M* and through a neighboring point *M′* on the curve is called a secant (Fig. 18). As *M′* approaches *M*, line *MM′* rotates around *M*, and in general, at the limit, when *M′* merges with *M*, this line merges with a straight line *MT* which is said to be tangent to the curve at *M*.

If now we consider two curves *C* and *C′* that have a common point *M*, the tangents *MT* to *C* and *MT′* to *C′* define a plane and form within this plane an angle *A* which is known as the angle of the two curves at *M*. When this angle is a right angle, we say that the curves are orthogonal at *M* (or perpendicular, in the case of two straight lines). In Euclidean geometry, from a given point to a given straight line one can draw one perpendicular line and only one.

Let us note that if the angle formed by two segments is *A*, since one causes *OD* to rotate in order to make it coincide with *OD′*, this angle is also *A* increased by any number of turns. It is also equal to a whole turn diminished by *A*, if one causes *OD′* to rotate in order to make it coincide with *OD*.

b3. DISPLACEMENTS; EQUAL FIGURES. A figure is formed by a number of line segments. For example, a triangle such as *ABC* is formed of three segments of straight lines, *AB*, *BC*, and *CA* (Fig. 19). Circles or arcs of circles form figures. Thus semicircle *AB* and line segment *BC* form a figure. If all the lines that form the figure are in the same plane, the figure is said to be plane.

Given these assumptions, we can now define pointwise transformations of figures. From a figure *F* one can derive a cor-

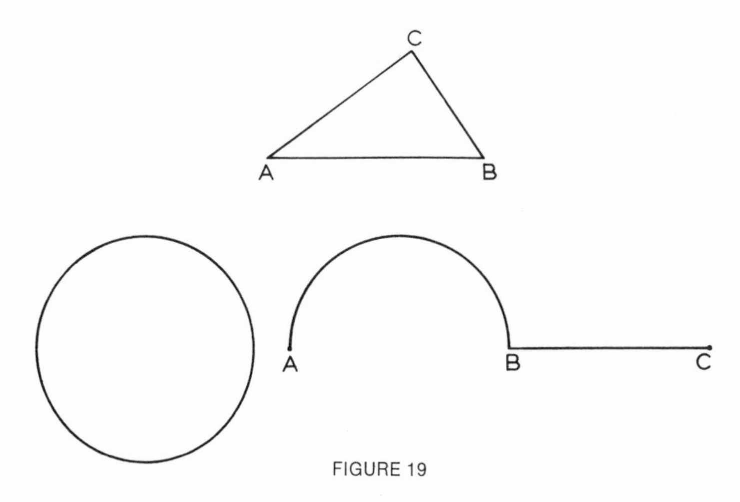

FIGURE 19

responding figure F' by transforming, according to a law given in advance, each point M of F into a point M' (Fig. 20). In this case all the points M' taken together will constitute figure F'.

For example, for every point M of figure F let us place a corresponding point M' in such a way that all straight lines MM' are parallel and that all lengths MM' are equal and in the same direction. It is said that F' is derived from F by a translation of

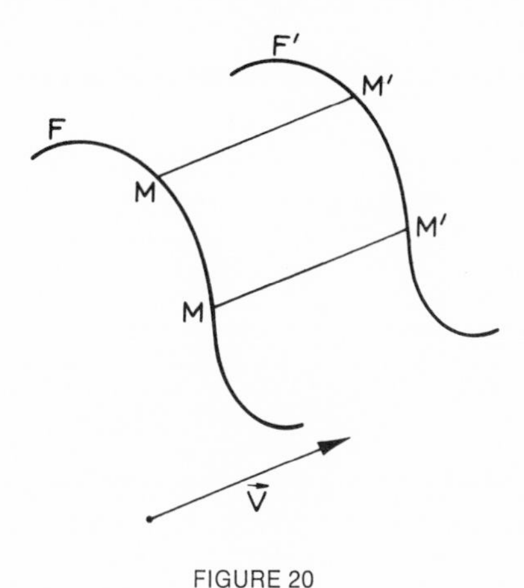

FIGURE 20

vector $\vec{V}$. This vector $\vec{V}$ is a segment of a straight line whose length is equal to MM' and whose direction is that of any straight line parallel to all straight lines MM'.

Let us now study another transformation. Given a straight line D and an angle α; a point M of F and the straight line D define a plane (Fig. 21). In this plane, one can draw one perpendicular and only one to D, which will be called MH. In the plane drawn at H perpendicularly to D, if we rotate line HM by angle α, M comes to M'. By applying this transformation to all the points M of F, one obtains a figure F'. The transformation by which one passes from F to F' is called the rotation around D of angle α.

Translations and rotations are particular kinds of transformations called displacements. Given now two figures F and F', we say that they are congruent if one can be derived from the other by a certain number of displacements.

In the plane, the Euclidean space of two dimensions, one can also determine translations and rotations of angle α around a point O of the plane. However, in this case, it may happen that two figures are congruent, if one regards them as being in a space of three dimensions, and yet in the plane on which they find themselves one cannot be derived from the other by means of displacements limited to this plane. This is the case of the two hands. They are congruent and yet, if they are laid flat on a

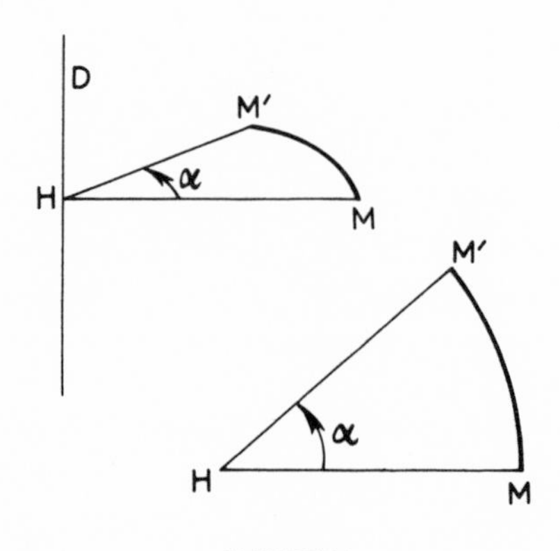

FIGURE 21

table, one cannot make them coincide. It is necessary that one of them leave the plane or, as we say, that we perform a reversal.

b4. HOMOLOGOUS FIGURES; SIMILAR FIGURES. Among the numerous pointwise transformations that one can imagine, there is one that has a fundamental importance for interpreting the observation of the universe, and that is homology.

Let us take a point O and a number K. Let us extend each straight line OM, M being a point on figure F, to point M' so that $OM' = K$ (Fig. 22). All such points M' constitute a figure F' which is said to be the homolog of F in a homology whose center is O and whose ratio is K.

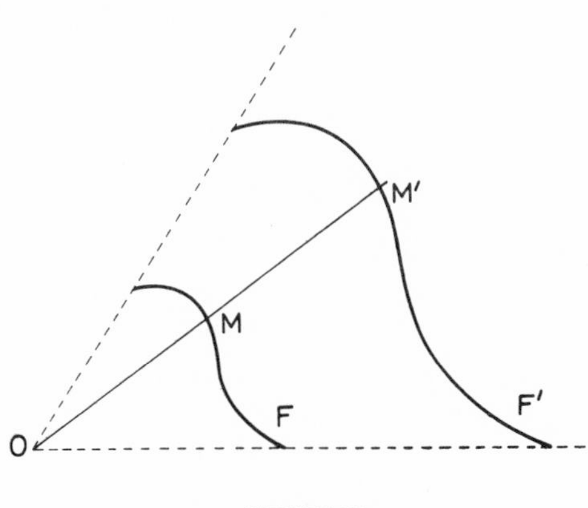

FIGURE 22

If, next, one performs certain displacements on F' which transform it into F'', F and F'' will be said to be similar. Similar figures are figures which are identical to the eye, but one of which is larger than the other.

The fact that the light ray is identified with the straight line of Euclidean space gives homology a fundamental role in our measurement of the universe. Let us consider a distant object AB, for example a line segment of unknown length, and another segment $A'B'$, parallel to AB, of known length (Fig. 23). If the eye of observer O sees A' merged with A and B' merged

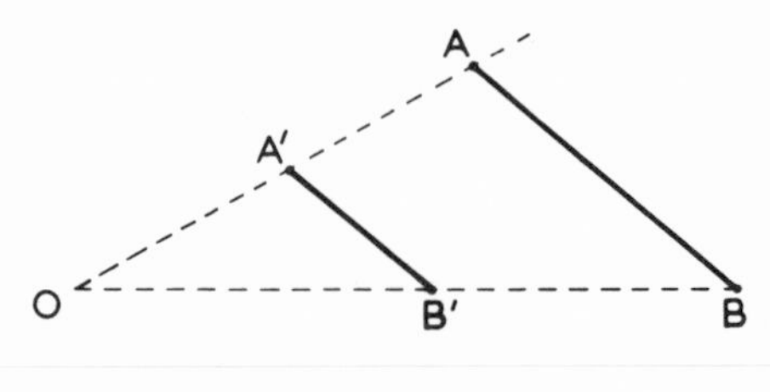

FIGURE 23

with B, one can draw a figure. $A'B'$ and AB may be derived from one another by homology, and hence

$$\frac{OA'}{OA} = \frac{A'B'}{AB} = \frac{OB'}{OB}$$

If one knows OA' and OA, one can then know the length of AB.

Now let there be a point A far away in the universe. Two observers O and O' note directions OA and $O'A$ and consequently angles $O'OA$ and $OO'A$; they also measure the distance between them OO' (Fig. 24). They can then draw on a sheet of paper a triangle $OO'A$ which will be similar to the real triangle $OO'A$ and thus determine distances OA and $O'A$.

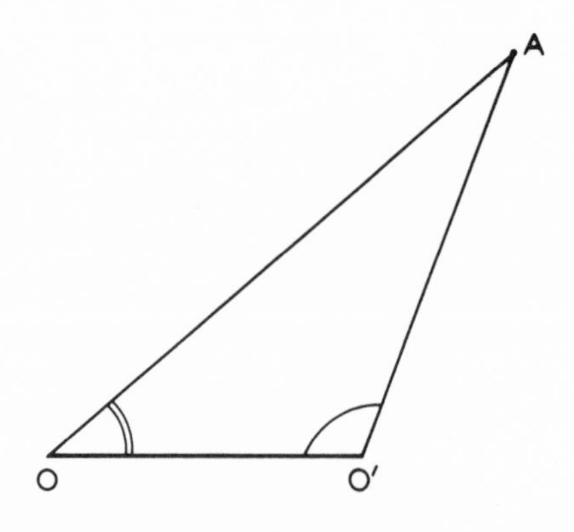

FIGURE 24

Thus Euclidean geometry offers the possibility of measuring the distant universe without direct contact; only the difficulties of observation and the precise measurement of angles limit the precision of this method.

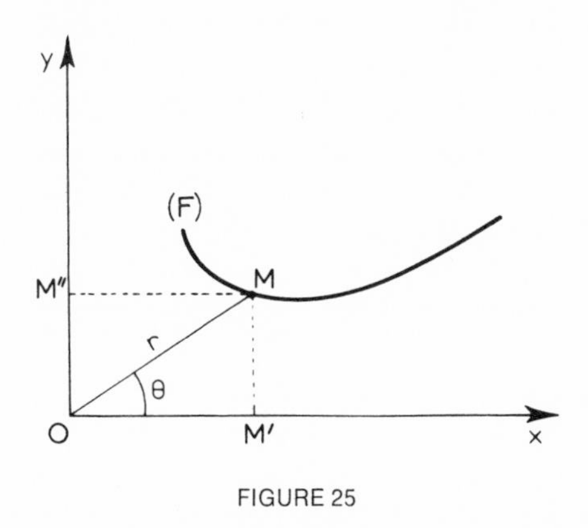

FIGURE 25

c1. ANALYTIC GEOMETRY. On a plane, let us consider two directional axes Ox and Oy which intersect at O and are perpendicular (Fig. 25).

Let us project a point M on the plane to M' on Ox and to M'' on Oy; the algebraic symbols x and y which measure the respective lengths of OM' and OM'' on Ox and Oy are called the coordinates of M. Knowing these two numbers, the abscissa x and the ordinate y, determines point M in a unique way. One could also determine the position of M by using the number r which measures the length of OM and the angle θ formed by OM with axis Ox. In any case, since a plane is a space of two dimensions, the position of a point M will be determined by the knowledge of two numbers.

If now M describes a curve F, the coordinates of M are not arbitrary but are bound to each other by a certain relationship:

$$f(x,y) = 0 \quad \text{or} \quad g(r,\theta) = 0$$

Each of these relationships is called the equation of curve F.

The same relationships can also be expressed by saying that x and y are functions of a single parameter u:

$$x = \phi(u)$$
$$y = \psi(u)$$

Points which belong to two curves, curve F whose equation is $f(x,y) = 0$ and curve F' whose equation is $f'(x,y) = 0$, have coordinates which verify both of these equations, and the geometric problem of determining the points of intersection of the curves is reduced to the algebraic problem of solving two equations with two unknowns.

Similarly, let us suppose that F' is derived from F by a homolog whose center is O and whose ratio is K. If x' and y' are the coordinates of M', it follows that

$$x' = Kx$$
$$y' = Ky$$

and since the equation of F is $f(x,y) = 0$, the equation of F' will be

$$f\left(\frac{x'}{k}, \frac{y'}{k}\right) = 0$$

This also shows us what will be the equation of a straight line passing through O (Fig. 26). On such a straight line, the

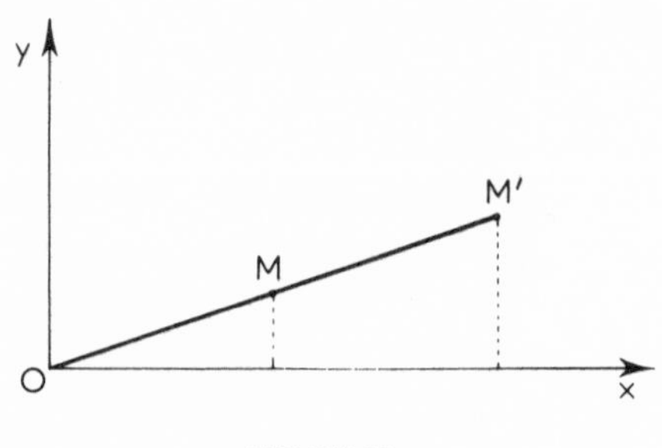

FIGURE 26

ratio y/x will be constant, no matter where point M is. Let a stand for this ratio. This gives us

$$\frac{y}{x} = a \quad \text{or} \quad y = ax$$

If the straight line does not pass through O it is easy to see that the equation is

$$y = ax + b, \text{ if } b \text{ is a constant}$$

Let us try to find the distance ds of two neighboring points, one point M whose coordinates are x and y, and the other point M' whose coordinates are $x + dx$ and $y + dy$, dx and dy being small quantities (Fig. 27). If we apply the Pythagorean theorem

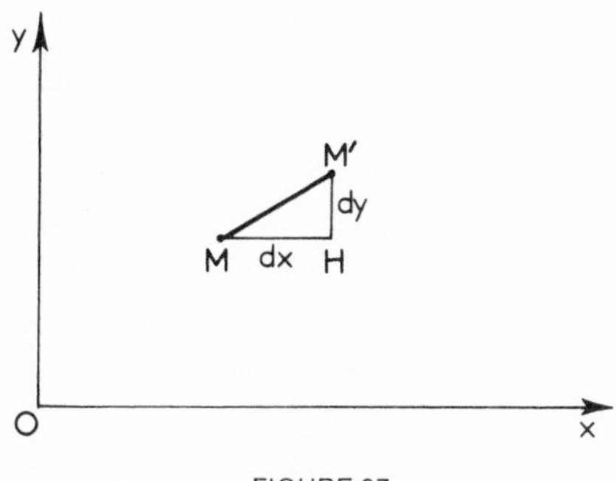

FIGURE 27

to triangle $MM'H$, in which MH is parallel to Ox and $M'H$ is parallel to Oy, we immediately get

$$\overline{MM'}^2 = \overline{MH}^2 + \overline{M'H}^2$$

or

$$\overline{ds}^2 = \overline{dx}^2 + \overline{dy}^2$$

If the points are not near one another, the above formula still gives the distance between M and M', that is, the length of

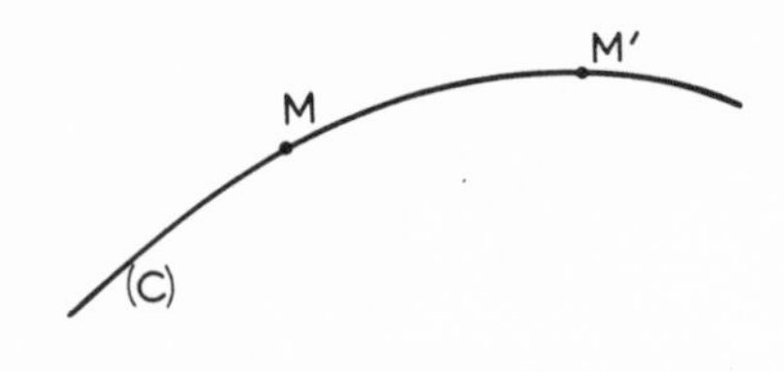

FIGURE 28

line segment MM'. If now one wants to determine the length of the segment of curve (C) between M and M', one must proceed differently. Let us express the equation of the curve (Fig. 28) in the form

$$x = \phi(u)$$
$$y = \psi(u)$$

To determine the value of u_1, let us assume that the x and y coordinates above of u are those of M, and to determine the value of u_2 let us use those of M'. Let us start at point M and place side by side along (C) a series of small rules whose length is ds. The sum of the length ds placed end to end from M to M' will be the length of MM' on (C). This is represented by

$$\int_{u_1}^{u_2} ds$$

which is read: the integral of ds from u_1 to u_2. When one increases u by a small quantity du, dx and dy increase by small quantities dx and dy, and

$$dx = \phi(u)du$$
$$dy = \psi'(u)du$$

$\phi'(u)$ and $\psi'(u)$ being, by definition, the derivatives, with respect to u, of $\phi(u)$ and $\psi(u)$. Therefore,

$$\int_{u_1}^{u_2} ds = \int_{u_1}^{u_2} \sqrt{dx^2 + dy^2} = \int_{u_1}^{u_2} \sqrt{\phi'^2(u) + \psi'^2(u)}\, du$$

If now one imagines a Euclidean space of three dimensions, this space will be expressed by three axes each of which is perpendicular to the other two, Ox, Oy, and Oz (Fig. 29). In

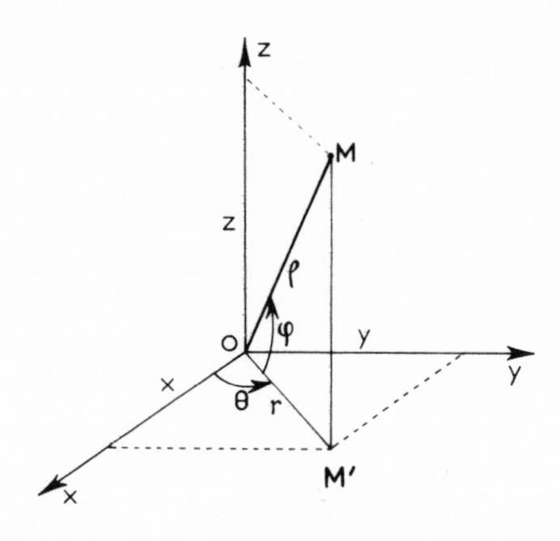

FIGURE 29

projecting a point M onto the plane defined by Ox and Oy, if x and y are the coordinates of M' on this plane, M can be exactly determined if one knows x, y, and dimension z above plane Ox, Oy. One can also work from r and θ of point M' and dimension z (cylindrical coordinates); or from distance OM, called ρ, angle of lines OM', OM, called ϕ (the latitude of M), and angle θ. These three variables taken together are known as spherical coordinates. In any case, since space has three dimensions, one needs three parameters in order to locate M.

An equation such as $f(x,y,z) = 0$ defines a two-dimensional surface surrounded by a three-dimensional space. It is easy to demonstrate that, no matter what the constant numbers a, b, c, and d are, the equation $ax + by + cz + d = 0$ is the equation of a plane.

If M is located on a sphere whose center is O and whose radius is R, distance OM is equal to R and $R^2 = \overline{OM}^2$, so the equation of this sphere will be

$$x^2 + y^2 + z^2 = R^2$$

One can likewise express ds^2, or the square of the elementary distance, as

$$\overline{ds}^2 = \overline{dx}^2 + \overline{dy}^2 + \overline{dz}^2$$

which gives the distance between two neighboring points. If a point M is displaced on a surface, its coordinates are not independent, they are bound by the relationship that gives the equation of the surface. One can also say that these coordinates depend on two parameters u and v (since M is then in a space of two dimensions), and the $\overline{ds}^2$ will be expressed as a function of du and dv.

II. TIME

a1. FUNDAMENTAL DEFINITION. In classical physics and mechanics, time is a fundamental dimension together with length and mass.

However, its immaterial aspect renders it inaccessible to the measurement necessary to a mathematical formulation of physical laws, except in an indirect and very particular way.

In fact, time appears in equations as a parameter, t, with no implication about the nature of this parameter except that it is related to a vague and intuitive notion of time stretching from minus infinity to plus infinity.

Let us suppose that a quantity x that is capable of being measured—a length, for example—has been related to this parameter t by a theory derived from a physical law and by experimental data. This relationship may be expressed, for example, by the functions

$$x = f(t)$$

or

$$t = g(x)$$

It is said that quantity x changes in time, these changes being expressed by the first formula. If now, at various instants, in the intuitive sense of this word, one takes measurements x_1, x_2, etc., of the quantity x, the second formula enables one to associate with these measurements certain values t_1, t_2, etc., of the variable t. One now says that one has established a scale of time. The phenomenon with which the measurable quantity x is associated constitutes a clock, and the parameter t, thus defined, is called Newtonian time.

Clearly, it is necessary that another clock, the law of whose movement may be expressed, for example, by the equation $y = h(t)$, provide values of t which are identical to those given by the clock $x = f(t)$ for the same instants. It is the principles of dynamics, the theories of motion that are derived from them, and the accuracy of the determination of the constants that enter into the formulas and are furnished by observation, that provide this guarantee. The observation of a discrepancy in the information given by two clocks (for example, the center of gravity of the earth in its movement around the sun and the internal vibrations of a molecule of ammonia) obliges us to carefully verify the hypotheses formed in the theory of each of the clocks and the results of observation. Conceivably, after all these verifications have been made, we may be obliged to call into question some of the principles that have led to the theories of these clocks.

In particular, we shall see further on that two clocks which are identical when they are at rest in relation to one another cease to define the same scale of time when they are in motion in relation to one another. (In order to bring out this fact, however, it is necessary that the relative speeds of the two clocks be very high.) This fact will lead us to reject the absolute character of Newtonian time, which is not called into doubt in classical mechanics.

b1. SIDEREAL TIME. If one regards the stars as stationary on the sphere of fixed stars, that is, if one overlooks their own movements, and if one assumes the earth to be spherical and homogenous, one can construct a very simple theory about the earth's motion around its center of gravity. In effect, in this case, the forces of gravitation (the attraction of the sun and the planets) have a zero torque about the center of gravity of the earth, and mechanics teaches us that the earth must therefore have a motion of uniform rotation around a fixed axis (its own). In other words, the hour angle of a given star will be expressed by the formula

$$H = At + B$$

if A and B are constants. If one chooses the starting instant of the scale and the unit in which t will be measured in such a way that B is zero and A is equal to 1, we will have

$$H = t$$

Thus the rotation of the earth defines a scale of time. This theory is only approximate, but it is close enough so that the ancients, for want of accurate enough observation, did in fact identify time with the rotation of the sphere of the fixed stars.

In fact, the "guiding star" chosen is the γ point, and the hour angle T of this point in a given place is known as the true sidereal time of that place. Hipparchus had already observed that the vernal point is not fixed on the sphere of the fixed stars. We now know how to demonstrate mathematically that this must be so because of the action of the sun and moon on the equatorial bulge of the earth, which causes shifts of the vernal point known as precession and nutation. In a given place (Greenwich, for example), the theory provides the following formula for true sidereal time:

$$T = At + B + Ct^2 + N$$

if A, B, and C are constants and N is the sum of a large number of periodical functions of t involving the longitude of the sun and the moon, and other quantities. All the terms of N put together are known as the nutation. The quantity

$$T_m = T - N$$

is called the mean sidereal time of the place in question. This is not a linear function of t because of the presence of the so-called term of precession, Ct^2. In any case, the formulas for T and T_m define two scales of time.

However, we now have good reason to believe that these scales of time are very imperfect. We already know three phenomena which make the use of sidereal time over very long periods doubtful:

▪ The slowing down of the rotation of the earth owing to the friction of the tides. The length of a day is supposedly increasing, at the present time, by one second per 500 centuries, but this effect is no doubt not linear.

▪ Seasonal variations in the rotation of the earth, probably caused by significant displacements of masses in the atmosphere.

▪ Sudden variations, of a random character, which seem related to solar activity.

b2. TRUE SOLAR TIME AND MEAN SOLAR TIME. In a given place (Greenwich, for example), the hour angle of the center of the sun is, by definition, the true solar time of that place. As we have seen in Part I, this angle H is expressed as

$$H = T - \alpha$$

when T is true sidereal time and α is the right ascension of the sun. The theory provides the expressions T and α in terms of Newtonian time t, so that we have:

$$H = H_0 + H_1 t - E$$

H_0 and H_1 being constants. E is called the equation of time, and represents a sum of periodical functions as follows:

$$E = C + R - L$$

Here C is the "equation to the center," reflecting the fact that the movement of the earth around the sun is not a circular and uniform movement. R is the "reduction to the equator," and reflects the fact that the equator and the ecliptic are not identical but form an angle of 23°30′. L represents everything that is not a linear function of t in true sidereal time (that is, $Ct^2 + N$).

Mean solar time is by definition

$$H_m = H + E = H_0 + H_1 t$$

The scale of time so defined regulates daily life. Let us note that the fact that average solar time is a linear function of t facilitates the construction of man-made clocks which will be responsible for reproducing this scale of time in practice (pendulum clocks, watches, etc.).

b3. EPHEMERIS. Mean solar time certainly does not provide an excellent representation of Newtonian time, since by definition it must contain the unforeseeable inequalities of sidereal time which we have mentioned. The difficulty of establishing a scale of Newtonian time based on the earth's rotation (sidereal time) has to do with the shortcomings of the clock, a precise theory of which it is difficult to construct.

As for the clock that defines mean solar time, it is in a sense made of two "parts": the sun, about which the theory is well known, but also the earth, so that one falls back into the foregoing difficulties.

The scale of ephemeris time tries to free itself from the movement of the earth's rotation and is defined solely in terms of the earth's center of gravity in its movement of translation around the sun (or, what amounts to the same thing, the sun's center of gravity in its movement around the earth).

All the forces of tides caused by the moon and aleatory effects on the movement of rotation become negligible, and the clock then consists of a material point moving as a result of universal gravitation in the field of the sun. Of course, it will be necessary to take into consideration the disturbances of the other planets, something which celestial mechanics enables us to do.

Newcomb's theory gives the following formula for the longitude of the sun:

$$L = L_0 + L_1 t + L_2 t + P$$

L_0, L_1, and L_2 being constants and P the sum of certain periodical terms.

The formula

$$L_s = L_0 + L_1t + L_2t^2$$

thus defines the scale of ephemeris time. One need only know L_s for a given moment in order to know t. In fact, the "sun clock" is so slow (one whole turn around the dial in a year) that we use the movement of the moon as an intermediary in order to calculate ephemeris time.

III. MECHANICS

The object of mechanics is the study of bodies in motion and its principal problem is to determine movement when one knows the causes that are producing it and the situation of the system (positions and velocities) at a given moment, known as the initial instant.

Mechanics utilizes the notion of the material point, an idealization, as in geometry, in which any system is regarded as a group of such points that may be finite or infinite in number. We shall speak here only of systems each of which contains a single point, and which constitutes an adequate approximation in many cases in astronomy, where we tend to identify the planets with their centers in studying their movements around the sun. We proceed similarly in studying the movements of the stars in relation to one another.

a. KINEMATICS.

a1. GENERAL PRINCIPLES. This branch of mechanics describes movements without concerning itself with their causes. The framework of classical kinematics is provided by Euclidean analytic geometry. Let M be a point, and x, y, and z its coordinates in relation to a system of Cartesian axes. In the course of time, x, y, and z vary and are therefore functions of the parameter t, which represents time. This gives us

$$x = f(t); \quad y = g(t); \quad z = h(t)$$

One can simplify the above equations by considering vector $\overrightarrow{OM}$, O being the point of origin of the coordinates, and by writing

$$\overrightarrow{OM} = \overrightarrow{U}(t)$$

This equation expresses the fact that vector $\overrightarrow{OM}$ is a vectorial function of the parameter t.

When time t varies, the furthest point M of vector $\overrightarrow{OM}$ describes a curve which is defined parametrically by any of these equations, and which is known as the trajectory of M.

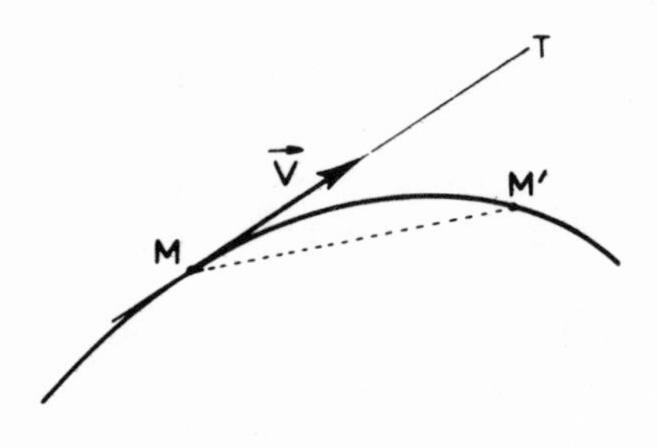

FIGURE 30

a2. VELOCITY AND ACCELERATION. At time t, the moving body is at point M; at the end of a very short interval $t + \Delta t$ (Δt is very small), it is at point M' (Fig. 30):

$$\overrightarrow{OM'} = \overrightarrow{U}(t + \Delta t)$$

As Δt tends toward zero, point M' comes closer to M, and the limiting position of line MM' is the tangent to the trajectory at point M.

We now have

$$\frac{\overrightarrow{OM'} - \overrightarrow{OM}}{\Delta t} = \frac{\overrightarrow{MM'}}{\Delta t} = \frac{\overrightarrow{U}(t + \Delta t) + \overrightarrow{U}(t)}{\Delta t}$$

When Δt tends toward zero, both the numerator and the denominator of the last fraction tend also toward zero, but their ratio may tend (and usually does in the cases studied by mechanics) toward a clearly defined vector $\vec{V}$, the vectorial derivative of $\vec{U}(t)$ at time t, which is called the vector velocity of M at time t. This vector $\vec{V}$ clearly lies along the straight line, the limiting position of MM' when M' approaches M, that is, by the tangent at M to the trajectory.

The components of vector $\vec{V}$ on axes X, Y, and Z are obtained by taking the derivatives of the second terms of the first group of equations in relation to t:

$$X = f'(t); \quad Y = g'(t); \quad Z = h'(t)$$

If the trajectory is a straight line (rectilinear motion), vector velocity always lies along the trajectory. If, in addition, this vector has a constant length, we say that the motion is rectilinear and uniform. The coordinates of M are then linear functions of time t:

$$x = at + b; \quad y = a_1t + b_1; \quad z = a_2t + b_2$$

If a, b, etc. are constants, we have

$$X = a; \quad Y = a_1; \quad Z = a_2$$

The length of vector velocity is called the velocity of the moving body at instant t. In the case of the rectilinear and uniform movement referred to above, this velocity v is equal to

$$v = \sqrt{a^2 + a_1{}^2 + a_2{}^2}$$

If, in the general case, one considers a point P which is such that vector $\vec{OP}$ is at every moment equal to vector $\vec{V}$—which means that it is parallel to vector $\vec{V}$, has the same length, and is in the same sense as $\vec{V}$—one has defined a point P which, when M describes its trajectory, describes a curve that is called the hodograph of the motion of M (Fig. 31).

If the motion of M is rectilinear, the hodograph is a straight line, since the dirèctrix of $\vec{V}$ is a fixed straight line, the trajectory.

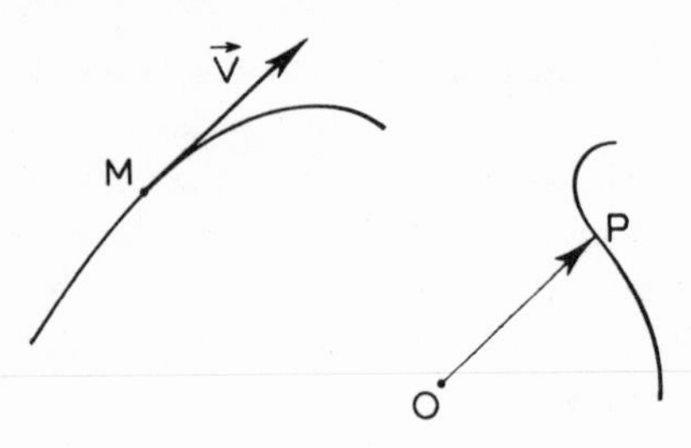

FIGURE 31

If the motion is not only rectilinear but uniform, the hodograph is reduced to a point, since vector velocity is always equal to itself.

In the case where the motion of the moving body M is undefined, the vector velocity $\vec{V}$ of M is clearly a function of time t. Let $t + \Delta t$ be a moment very close to t and let us consider the vector:

$$\frac{\vec{V}(t + \Delta t) - \vec{V}(t)}{\Delta t}$$

When Δt tends toward zero, the numerator and denominator of the fraction tend simultaneously toward zero, but the ratio may tend toward a clearly defined vector $\bar{\gamma}$. This vector is called the vector acceleration of point M. If we refer to the definition of the hodograph, we have no trouble seeing that vector $\bar{\gamma}$ is none other than the vector velocity of point P on the hodograph. Of course, the acceleration vector is not, generally speaking, tangent

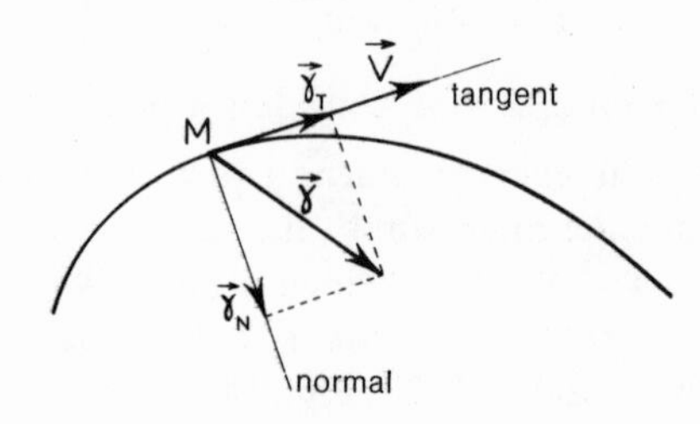

FIGURE 32

to the trajectory at point M. This vector can be broken down into two vectors of which it is the sum. The first, carried by the tangent at M, is called the tangential acceleration; the second, carried by the normal to the trajectory at M (that is, the perpendicular to the tangent to the trajectory at M), is called normal acceleration (Fig. 32).

It can be demonstrated that the magnitude of tangential acceleration $\vec{\gamma}_T$ is

$$\vec{\gamma}_T = v'(t)$$

$v'(t)$ being the derivative of velocity v in relation to time. We also have

$$\vec{\gamma}_N = \frac{v^2}{R}$$

$\vec{\gamma}_N$ being the magnitude of normal acceleration. R is a geometric quantity attached to point M of the trajectory curve which is called the radius of curvature at M. If the trajectory is a circle whose radius is R, the radius of curvature at any point on the circle is equal to this radius R. If the trajectory is a straight line, the radius of curvature is infinite at any point on the straight line.

▪ *Rectilinear and uniform movement.* The trajectory is a straight line, hence the acceleration vector is along this straight line. This is confirmed by the fact that $\vec{\gamma}_N$ is zero since R is infinitely large.

Moreover, v being constant, the derivative $v'(t)$ is zero. In other words, in the case of a rectilinear and uniform motion, the acceleration vector is zero.

▪ *Circular uniform motion.* Let us suppose that M is moving in a circle whose center is O and whose radius is R. The motion is uniform if the velocity vector is constant. If v is this quantity, its derivative is therefore zero, and we have (Fig. 33)

$$\vec{\gamma}_T = O; \ \vec{\gamma}_N = \frac{v^2}{R}$$

Since v and R are constant, $\vec{\gamma}_N$ is constant. The normal, at any

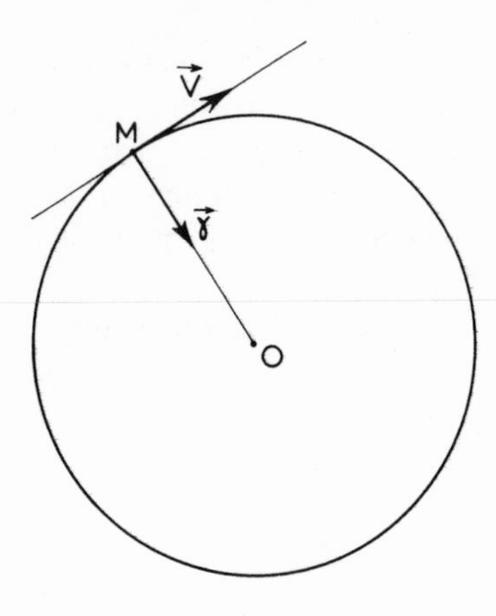

FIGURE 33

point *M* on the circle, passes through the center *O*, since we know that the tangent at a point *M* of a circle is perpendicular to *OM*. In circular and uniform motion, therefore, acceleration is a vector of constant magnitude along *OM*.

a3. RELATIVE MOVEMENTS. Given a reference defined by a system of Cartesian coordinates *Ox*, *Oy*, *Oz*, let us consider two moving points *A* and *M* (Fig. 34). One can determine the move-

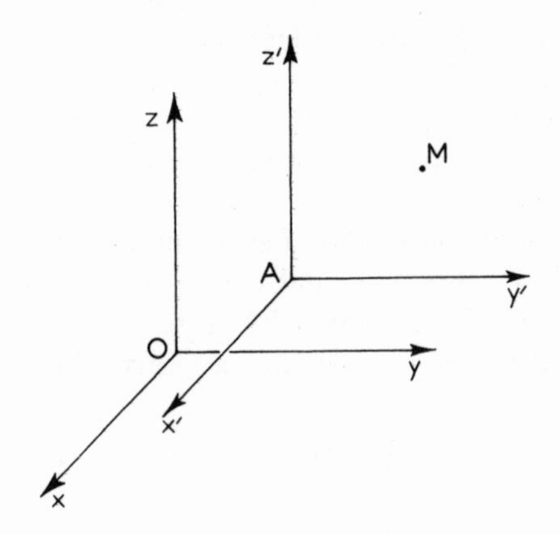

FIGURE 34

ment of M either in relation to the reference starting at O, or in relation to a reference starting at A whose axes Ax', Ay', Az' are parallel to axes Ox, Oy, Oz. We know, of course, that

$$\overrightarrow{OM} = \overrightarrow{OA} + \overrightarrow{AM}$$

according to the relationship of Michel Chasles. If we take the derivatives of the two members in relation to time t, we have

$$\overrightarrow{V_m} = \overrightarrow{V_a} + \overrightarrow{V_{am}}$$

In other words, the vector velocity of M in relation to O is equal to the sum of the vector velocity of A in relation to O and the vector velocity of M in relation to A. By taking the derivatives of the two members in relation to time, one would have an analogous result concerning acceleration.

b. DYNAMICS

b1. GENERAL PRINCIPLES. The purpose of dynamics is to define mathematically the causes of movements and, when these causes are given for a particular problem, to deduce from them the corresponding movement. In classical mechanics, it is necessary to make certain fundamental hypotheses which are generally confirmed by experiment.

▪ Absolute space has a direction. In other words, every material point may be said to be either at rest or in motion in relation to a reference which is itself regarded as absolutely fixed. This reference is called an absolute reference. Any point O in absolute space may be taken, therefore, as the origin of this reference; any fixed axes Ox, Oy, and Oz starting at O may be taken as axes of reference. For example, in studying movements in the solar system, the origin of the reference will be the center of the sun, and the directions of the axes will be the directions of distant stars. Naturally, the movement of a point may be related to any reference which is moving in relation to the absolute reference, provided one takes into account the formulas of relative motion.

▪ Since the mathematical causes of a movement are given for

each instant, the movement of a point may be determined in a unique manner if one knows the position and vector velocity of the point in question at a given instant. In order to determine the position and the velocity of a point at a given instant, one must know for that instant six constants: the three coordinates of the point and the three components of vector velocity. These constants are called the initial constants or initial conditions.

▪ Given a point *M* which is moving in relation to the absolute reference and $\vec{\gamma}$ its vector acceleration at instant *t*, there exist (1) a positive quantity *m* attached to point *M* and independent of the motion in question, called the inertial mass of *M* or simply the mass of *M*; and (2) a vector $\vec{F}$ originating at *M*, dependent on the position and velocity of *M* at instant *t* and sometimes on time *t* itself, which is such that

$$\vec{\tau} = \frac{\vec{F}}{m}$$

Vector $\vec{F}$ is called the force applied at *M*. Vector $\vec{F}$ indicates the presence of one or more other bodies acting on *M*. One can, in certain cases, make hypotheses about the size and direction of this force and thereby deduce the acceleration of *M*, as we shall see further on.

Let us take $\vec{F}$ and apply this force first to *M* and then to another point *M′* whose mass is *m′*. The acceleration of *M′* will thus be

$$\vec{\gamma}' = \frac{\vec{F}}{m'}$$

Therefore

$$\vec{\gamma}' = \vec{\gamma}\frac{m}{m'}$$

or, the accelerations of *M* and *M′* will be vectors having the same direction and the same sense, but different magnitudes. The larger m' is, the smaller is the size of $\vec{\gamma}'$; it would become zero if m' increased indefinitely, which enables us to understand intuitively the meaning of the term inertial mass.

Now let us suppose that the motion of M is rectilinear and uniform. We have seen that in this case $\vec{\gamma}$ is zero, hence $\vec{F}$ is also zero. One would also have the same result if M were at rest in relation to the absolute reference. (In this case, not only vector acceleration but also vector velocity would be zero.) Conversely, if one applies to a point M of mass m a force $\vec{F}$ which is constantly zero, the acceleration of M will be constantly zero and this point will either be at rest or will have a rectilinear and uniform motion. This property is known as the principle of inertia.

b2. GALILEAN REFERENCES. Let us consider the absolute reference $Oxyz$ and another reference originating at A whose axes are parallel to Ox, Oy, Oz, and let us assume that the motion of A in relation to the absolute reference is rectilinear and uniform. The acceleration of A is zero and consequently, in accordance with what we saw above,

$$\vec{\gamma}_{OM} = \vec{\gamma}_{AM}$$

or, the acceleration of a point M in relation to the absolute reference is identical to the acceleration of this point in relation to the reference originating at A. In particular, if the acceleration is zero for the motion in relation to the absolute reference, it is also zero for the motion and relation to the reference originating at A. If a force $\vec{F}$ is applied to M, of mass m, the fundamental equation

$$\vec{F} = m\vec{\gamma}$$

is valid as it stands for either reference, and there is no reason, therefore, to distinguish one from the other from the point of view of dynamics. Any reference determined by rectilinear and uniform translation in relation to the absolute reference is called a Galilean reference. Any Galilean reference may be chosen in order to study the motion of a point, the equation of the motion being the one given above. This property is known as the principle of Newtonian relativity.

Now let us suppose that we wish to study the motion of a point, and that we make the hypothesis that we know at every instant the force $\vec{F}$ that is acting on it. One can mathematically determine this motion by utilizing the above equation and by knowing the initial conditions (which we assume to be known here without any error). If the observed motion does not correspond to the calculated motion, three hypotheses may be considered: (1) The postulate of the existence of the absolute reference is not legitimate. In classical mechanics we disregard this possibility. (2) Force $\vec{F}$ is imperfectly known. (3) The reference used is not really Galilean.

We shall see that if one finds the last hypothesis to be correct, one can go back to hypothesis (2) by correcting force $\vec{F}$.

b3. FORCES OF INERTIA. Let us suppose that the reference originating at *A* has its axes parallel to those of the absolute reference, but that the motion of *A* is undefined and not rectilinear and uniform.

We have already seen that

$$\vec{\gamma}_{OM} = \vec{\gamma}_{OA} + \vec{\gamma}_{AM}$$

That is, the vector acceleration of *M* in its motion in relation to the absolute reference originating at *O* is equal to the sum of the vector acceleration of the motion of *A* in relation to the absolute reference and the vector acceleration of *M* in relation to the reference originating at *A*.

Let us suppose that a force $\vec{F}$ is applied at *M* and that the reference starting at *A* is believed to be Galilean. The equation of *M*'s motion will be written

$$\vec{F} = m\vec{\gamma}_{AM}$$

and we should also have

$$\vec{F} = m\vec{\gamma}_{OM}$$

since the two references are, in principle, indistinguishable. But the motion of M in relation to the reference starting at A will be completely different from the one we would obtain by solving one of the foregoing equations.

We shall have to write

$$\vec{\gamma}_{AM} = \frac{\vec{F}}{m} - \vec{\gamma}_{OA}$$

In other words, in the reference starting at A, point M is subjected to the force

$$\vec{F} - m\vec{\gamma}_{OA}$$

the sum of force $\vec{F}$ and of an imaginary force, $-m\vec{\gamma}_{OA}$, called the force of inertia.

This force of inertia is said to be imaginary inasmuch as it seems due not to the presence of bodies acting on the point but simply to the fact that the reference utilized has not been determined by rectilinear and uniform translation in relation to the absolute reference.

Let us take an example. In a train that is moving along a rectilinear track, we drop a ball. If the velocity of the train is constant, the motion of the reference consisting of the interior of the car is rectilinear and uniform. The ball that has been dropped is subject to a force $\vec{F}$ called gravity which has a constant value and which is directed downward. The ball falls, that is to say, it travels along a descending vertical with a motion that is uniformly accelerated, as everyone knows.

Now let us suppose that the ball is dropped while the train is accelerating, and that the observer has not been warned about this acceleration. The ball will no longer fall vertically, but will reach the floor of the car by traveling along a plane curve. The observer will find this result to be in contradiction to the law of falling bodies, unless he is aware that his reference, the car, is not a Galilean reference. He will predict the movement of the ball correctly if, independently of gravity, he takes into consideration the force of inertia, which will be represented here by a horizontal vector directed from the front of the train to the back of the train and equal to

$$-m\overrightarrow{\gamma_T}$$

m being the mass of the ball and γ_T the acceleration of the train. The sum Φ of these two vectors must be utilized in order to find the motion of the ball, and the formula will be, in relation to the car,

$$\overrightarrow{\Phi} = m\overrightarrow{\gamma}$$

NOTE: Up to now we have assumed that the axes of the reference originating at *A* remained parallel to the axes of the absolute reference. In this case we say that reference *A* is impelled by a movement of translation and that this is true no matter what the trajectory of *A* is. In general this is not the case; the system of axes of the reference proceeding from *A* may rotate around *A*. The formulas for the velocity of a point and its acceleration in relation to an absolute reference, when one knows its velocity and acceleration in relation to a moving reference and the motion of the moving reference, are more complex, therefore, than those we have given for those cases in which the moving reference is impelled by a movement of translation. It follows that the forces of inertia also have a more complex expression.

Let us take a rather simple example. The origin *A* of the

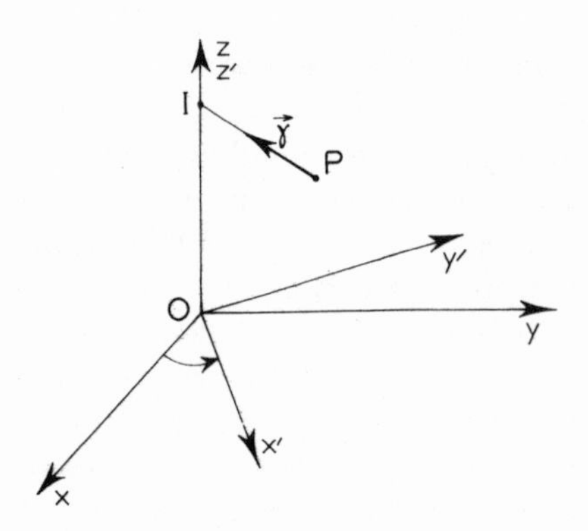

FIGURE 35

moving reference coincides with O (it is therefore fixed), and the axis Az' coincides with Oz, but axes Ax' and Ay' rotate around Oz with a uniform motion (Fig. 35). A point P connected to the moving reference, that is, fixed in relation to it, has a constant acceleration which is parallel to plane Ox, Oy and which intersects axis Oz at a fixed point I. Thus in studying the movements of the moving reference one must take into account forces of inertia parallel to plane Ox, Oy and opposing the accelerations of point P. These forces are called centrifugal forces.

Now let us suppose that a passenger is in a train which is moving with a uniform motion along a rectilinear track. If the train brakes suddenly, the passenger is thrown toward the front of the train. In effect, at the moment of braking, the reference frame constituted by the train is impelled by a movement which is still rectilinear but no longer uniform. Each point of this reference frame is subjected to an acceleration directed from the front of the train toward the rear, and consequently the passenger is subjected to a force of inertia which is proportional to his mass and to the acceleration of the train and which is directed from the back toward the front.

b4. MATERIAL SYSTEMS. The term "material system" refers to a system composed of a finite or infinite number of material points. Let us confine ourselves here to a system composed of two material points M_1 and M_2 whose masses are m_1 and m_2 (Fig. 36), and let us give a few definitions:

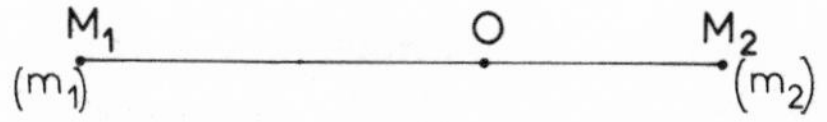

M_1 (m_1) — O — M_2 (m_2)

FIGURE 36

The term "center of mass" of the system (it is sometimes called "center of gravity") refers to a point O located on segment M_1M_2 in such a way that

$$m_1OM_1 = -m_2OM_2$$

The term "forces internal to the system" refers to those forces which the different points of the system exert on one another. Thus, if M_1 is the sun and M_2 the earth, we know that the sun attracts the earth; this attraction is an internal force. But one can also say that the earth attracts the sun. This constitutes a phenomenon known as the principle of action and reaction, which is stated as follows (for two material points): if M_1 exerts on M_2 a force $\vec{F}$, M_2 exerts on M_1 an equal and opposite force. In other words, in the system composed of M_1 and M_2, the sum of the internal forces is zero.

The term "external forces" refers to forces other than internal forces. Thus, the attraction exerted by the moon on the earth is a force external to the system consisting of the earth and the sun. Of course, as we have seen, external forces may be either given forces, or forces of inertia.

b5. THE LAW OF UNIVERSAL GRAVITATION. Newton formulated the following law: "Given two material points, M_1 with a mass m_1 and M_2 with a mass m_2, M_1 exerts on M_2 a force of attraction (which means that the vector force is directed from M_2 toward M_1) proportional to the product of m_1 and m_2 and inversely proportional to the square of the distance between M_1 and M_2."

This law is derived from sensory perception rather than controlled experiment. In fact, when it is used in conjunction with the fundamental equation

$$\vec{F} = m\vec{\gamma}$$

it enables one to determine the acceleration of a body and hence its motion, if one knows the initial conditions. The movements of the stars, as well as the falling of bodies and the phenomena of the tides, are perfectly explained by this simple law, which we shall, therefore, have every reason to think correct as long as experience or the precision of our measurements does not contradict it.

Of course, according to the principle of action and reaction, M_2 exerts an equal force of attraction on M_1. We shall see what the movements of M_1 and M_2 are under the effect of these forces (see Part V, *Astronomy*).

It can be demonstrated that a material system composed of an infinity of material points which has the form of a homogenous sphere may be identified with a material point located in the center of the sphere whose mass is the mass of the sphere.

Let us consider the earth. It attracts bodies located in the vicinity of its surface, according to Newton's law. The quantity of this force is GmM/R^2 if m is the mass of the attracted body, M the mass of the earth, G a constant (called the constant of universal gravitation), and R the radius of the earth. If $\vec{g}$ is the acceleration acquired by a body M of mass m, vector $\vec{g}$ is directed toward the center of the earth and its quantity is

$$\vec{g} = \frac{GM}{R^2}$$

It is therefore constant, and is called the acceleration of gravity.

IV. OPTICS

Optics deals with the properties of light, and experimentation led the scientists of the classical era to think that this science should be conceived of from two different points of view. In fact, certain phenomena suggest that light energy is propagated along certain directions called light rays, which are identified with the straight lines of Euclidean geometry. Thus we speak of the rectilinear propagation of light.

Other phenomena, which lend themselves to simple elementary verification, lead one to conceive of light as a vibratory phenomenon which is propagated from one point to another in a medium. The speed of propagation of the wave in a given medium is identified with the speed of a "luminous particle" along a light ray, so that the expression "speed of light" has meaning, no matter which hypothesis one accepts about the nature of light.

a. PHYSICAL OPTICS

a1. VIBRATORY PHENOMENA. Let us suppose that at a point M in space, a certain quantity varies in a periodical fashion. For example, let us suppose that this quantity, whether it is scalar or vectorial, oscillates in amplitude between the values $-a$ and $+a$, passing through the value 0, in a period of time T, and that the phenomenon repeats itself indefinitely: we say that point M is the locus of a vibratory phenomenon whose period is T, whose amplitude is a, and whose average is zero. If x is the value of the quantity in question at a given instant t, a particularly simple vibratory phenomenon will be one of which it will be true that

$$x = a \sin 2\pi \frac{t}{T}$$

When t increases by T, x acquires the same value.

We say in this case that the vibration at M is sinusoidal.

If now M is the locus of another vibratory phenomenon which is such that a quantity y is related to time by the formula

$$y = a \sin\left(2\pi \frac{t}{T} + \phi\right)$$

in which ϕ is a constant, we say that M is the locus of two vibratory phenomena of the same amplitude a and the same period T, which are not, however, in phase. The difference in phase is measured by the angle ϕ.

a2. PROPAGATION OF WAVES. Let us imagine that in a certain medium a point O is the locus of a vibratory phenomenon which we shall assume to be sinusoidal for the sake of simplicity. We have

$$x = a \sin 2\pi \frac{t}{T}$$

If the medium is such that the vibratory phenomenon may gradually spread to all points of the medium, we say that this medium is the locus of a wave originating at O. The speed v with which the phenomenon is propagated from starting point O will be characteristic of the medium; we call it the speed of propagation of the wave. If this speed is constant for every point in the medium and in any direction, we say that the medium is homogenous and isotropic.

Now let us consider a point M situated at distance d from O. This point will be reached by the vibratory phenomenon, which we assume to begin at O at time zero, after an interval

$$t = \frac{d}{v}$$

if v is the speed of propagation of the wave. Thus, if y represents the quantity that vibrates at point M, the equation of its variation will be identical to the equation of the variation of x at O, with a difference of d/v in the time, giving us

$$y = a \sin \frac{2\pi}{T}\left(t - \frac{d}{v}\right)$$

which may also be written:

$$y = a \sin\left(\frac{2\pi t}{T} - \frac{2\pi d}{vT}\right)$$

In other words, point M is the locus of a vibratory phenomenon identical to the one at O, with a difference in phase:

$$\phi = -2\pi \frac{d}{vT}$$

The product vT, often designated by λ, is known as the wavelength of the vibratory phenomenon; it is the distance traveled by the wave during one period. We find also that all points situated on the same sphere whose center is O and whose radius is d are in the same vibratory state at the same moment. This sphere constitutes what is called a wave surface.

a3. LIGHT IS A VIBRATORY PHENOMENON; INTERFERENCE. The fact that light is caused by a vibratory phenomenon can be proved by some rather simple experiments, notably Young's experiment with

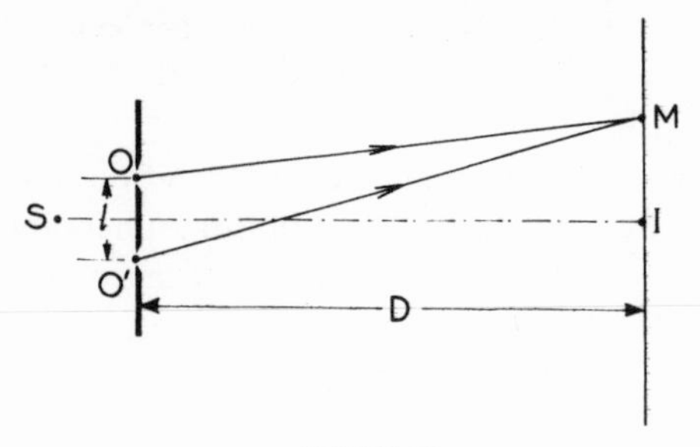

FIGURE 37

interference bands created by holes, which we shall describe (Fig. 37). Let us imagine two pointwise light sources that are in phase; for example, two small holes *O* and *O′* made in a screen and illuminated by the same source *S*. Now let there be another screen at distance *D* from holes *O* and *O′*. At point *I*, the waves that have left *O* and *O′* at the initial instant zero arrive at the same moment, since $OI = O'I$. *I* is the locus of a vibratory phenomenon which is the sum of the phenomena originating at *O* and *O′* which have, therefore, the same difference of phase as sources *O* and *O′*. Point *I* will be a brilliant point illuminated by the two sources. On the other hand, let us consider a point *M*, distinct from *I*; since distances *OM* and *O′M* are no longer equal, it is possible that the differences in phase of the two vibratory phenomena which are superimposed at *M*, with their original

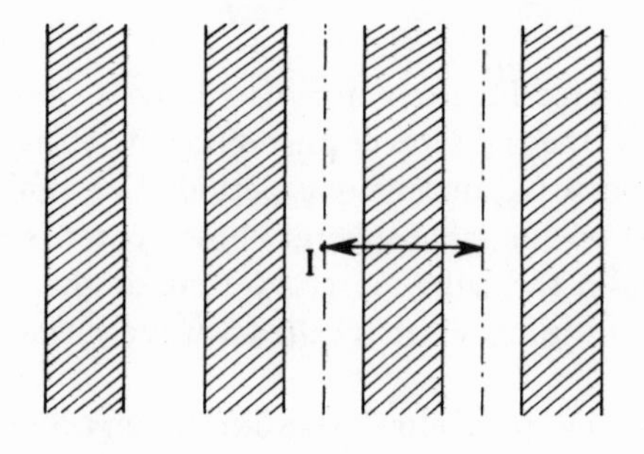

FIGURE 38

sources, are such that the amplitude of one is always equal and opposite to the amplitude of the other. In this case, point M will be constantly unilluminated. Finally, if the screen is far enough away from OO', we will observe on this screen a succession of alternately black and brilliant rings, called interference fringes (Fig. 38). It has been proved that the distance between the centers of two successive brilliant fringes is equal to

$$i = \lambda \frac{D}{l}$$

In this formula, D is the distance of the screen from sources O' and O, l is the distance between O' and O, and λ the wavelength of the light. Measurement of l, D, and i enables one to calculate λ. The illustration is a slight simplification of the actual pattern.

a4. THE SPEED OF LIGHT. One often designates as c the speed of light in a vacuum, that is, the speed of propagation of light waves in a vacuum. We accept as a fact of experience that this speed is constant and independent of the speed of the light source. The speed of light in a vacuum is equal to approximately 300,000 kilometers per second. It was measured for the first time in the seventeenth century, in the Paris observatory, by the Dane Olaus Roemer, who correctly attributed an apparent delay in certain phenomena involving the satellites of Jupiter to the time required by light to reach the observer.

In a medium which is not a vacuum, the speed v of light is

$$v = \frac{c}{n}$$

n being a number greater than one, known as the index of refraction of the medium. For example, the index of air at zero degrees is 1.0002926; for modern glass, this index is equal to about 1.5. This index varies with the wavelength of the light in question, which is always extremely short, varying from 4.1×10^{-7} meter for violet light (which may also be expressed as 0.41 microns) to 6.5×10^{-7} meter for red light. The vibratory phenomenon which produces light does extend beyond these two limits to include the areas of so-called ultraviolet and infrared, but the phenomena are no longer perceptible to the human eye.

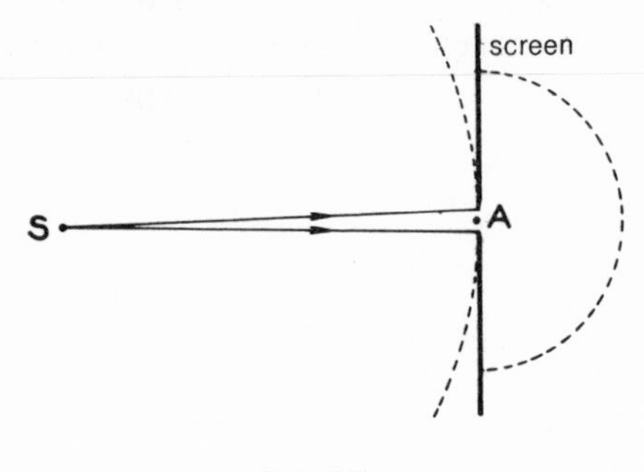

FIGURE 39

a5. THE PHENOMENA OF DIFFRACTION. Let us consider a closed surface containing a pointwise light source *S*. Huygens' principle teaches us that the vibrations propagated outside the surface are the same as if source *S* were replaced by certain sources located on the surface. Let us consider a screen which contains a hole *A* and which is tangent at *A* to a spherical wave surface whose center is *S* (Fig. 39). Point *A* becomes itself a source of light vibrations whose wave surfaces are spheres whose centers are *A*, so that even if the screen is opaque, there will be light outside of the cone whose summit is *S* and whose base is the tiny opening *A*. This phenomenon is known as the phenomenon of diffraction. Fresnel has shown, however, that if the size of the opening at *A* is particularly large in relation to the wavelength of the light, everything proceeds as if the light energy were transported along rectilinear rays called light rays.

Similarly, if in front of source *S* one places an opaque screen which is large in relation to the wavelength of the light, all of the section of the cone whose summit is *S* and whose base is the screen which is to the right of the screen on the drawing is in shadow (Fig. 40). These phenomena justify the convenient hypothesis of geometric optics, which we shall study further on.

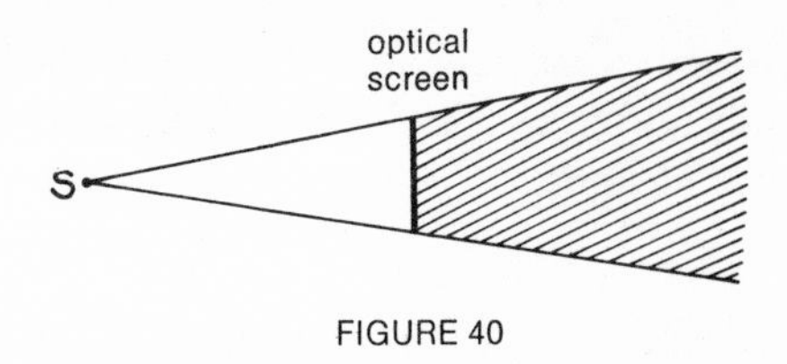

FIGURE 40

a6. POLARIZATION; THE NATURE OF LIGHT. Up to now we have talked about the vibratory nature of light without specifying the quantity x that was vibrating. The physicist Etienne Malus had noticed that when observing a light ray reflected first on one mirror and then on another, he could, by causing the second mirror to rotate around the first, arrive at a total extinction of the light ray. This phenomenon, known as the phenomenon of polarization, is too complex to be discussed in detail here. Let us simply note that it led to the conception that light is caused by the vibration of a vectorial quantity perpendicular to the direction of propagation.

Let V be this vectorial quantity; it can be broken down into two vectors $\vec{A}$ and $\vec{B}$ which are perpendicular to each other and to the direction of propagation. These two vectors $\vec{A}$ and $\vec{B}$ vary periodically with time; if these vibrations are in phase, the angle formed by $\vec{V}$ with vector $\vec{B}$ is constant: we say that the light is completely polarized. If $\vec{A}$ and $\vec{B}$ are not in phase, the light is partially polarized and the end of the vector describes an ellipse in the plane defined by $\vec{A}$ and $\vec{B}$ (Fig. 41).

It was James Clerk Maxwell who, in 1865, identified vectors $\vec{A}$ and $\vec{B}$ as an electrical field vector and a magnetic field vector respectively, and who thus originated the electromagnetic theory

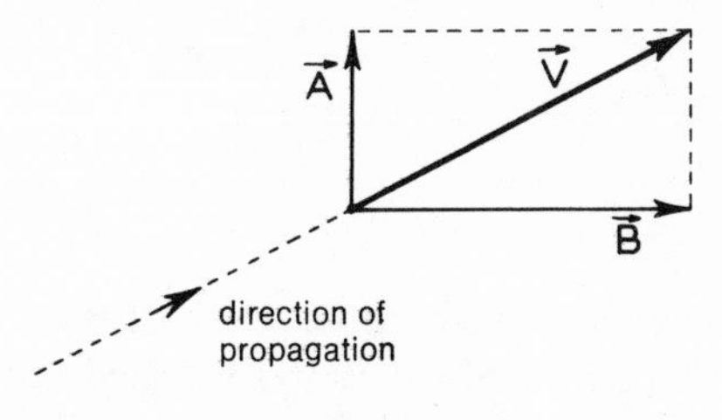

FIGURE 41

of light. This theory combined light phenomena with Hertzian waves, whose only difference is the wavelength, which is much shorter in the case of light.

b. GEOMETRIC OPTICS

In many cases, the hypothesis in which the light ray is identified with the straight line of Euclidean geometry, which we justified in (*a5*), is extremely useful. In particular, it lends itself admirably to explaining the formation of images in optical instruments.

b1. THE LAWS OF DESCARTES. Let us suppose that a surface *S* divides two media whose indices are n and n' respectively and let us consider a light ray *DM*. As it arrives in contact with surface *S* at point *M*, this light ray is subject to two phenomena:

▪ A phenomenon of reflection. Part of the light energy is sent back into the medium whose index is n along light ray *MD′*. If one describes as the normal the straight line at *M* perpendicular to the plane tangent at *M* to surface *S*, this normal and the straight lines *MD* and *MD′* are in the same plane, and the angles formed by *MD* and *MD′* with this normal are equal (Fig. 42).

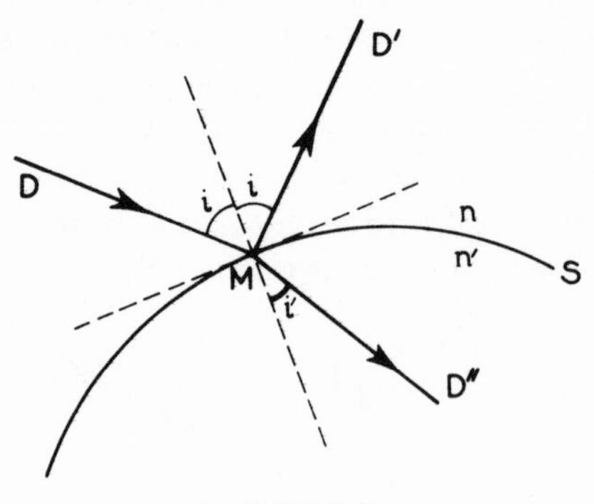

FIGURE 42

▪ A phenomenon of refraction. Part of the light energy travels through surface *S* and penetrates the medium whose index is n' following light ray *MD″*. *MD*, *MD″*, and the normal at *M* to the surface are in the same plane, and if i is the angle formed by MD and the normal, and i' the angle formed by MD'' and the normal, we have

$$n \sin i = n' \sin i'$$

Hence, if index n' is greater than n, i' is smaller than i, and the light ray, after refraction, approaches the normal. These laws concerning reflection and refraction are known as the Laws of Descartes.

b2. THE SLAB WITH PARALLEL SIDES; THE PRISM. A flat thin object with parallel sides—for example, a piece of glass—causes a light ray DM to undergo a first refraction which brings the refracted ray MM' toward the normal MN. But as it leaves the piece of glass at M', since normal $M'N'$ is parallel to MN, the emerging ray is, by reason of symmetry, parallel to ray DM (Fig. 43).

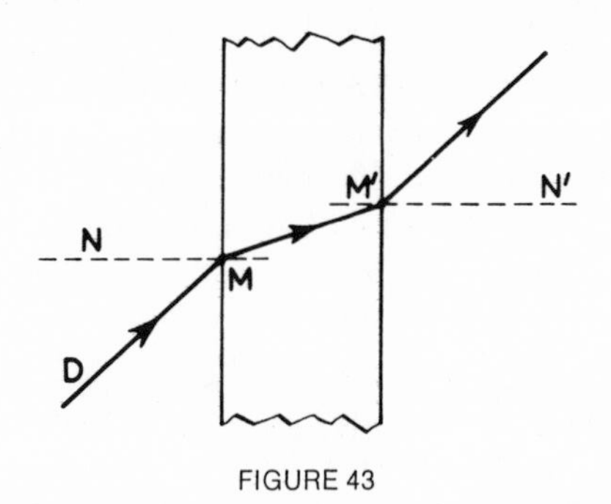

FIGURE 43

If now we consider a prism (Fig. 44), the drawing shows us at once that emerging ray $M'D'$ will not be parallel to incident ray DM. Moreover, since we have seen that the index of refraction depends on the wavelength of the light, if ray DM consists of two superimposed light rays of different wavelengths, there will be two emerging rays $M'D'$ and $M''D''$, which will not be combined. In particular, we know that if DM is a ray of white light, the prism will break it down as it emerges into a pencil of rays of all the colors of the rainbow, since color is the eye's way of in-

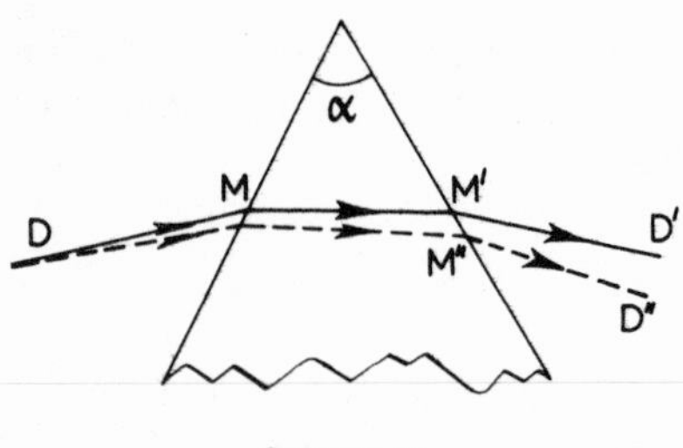

FIGURE 44

terpreting a given light wave. Later we shall see the significance of this possibility of analyzing the different components of a light ray.

b3. THE PARABOLIC MIRROR. A paraboloid of revolution is a figure obtained by rotating a parabola around its axis of symmetry. The paraboloid has an interesting geometric property: if any straight line *D* parallel to the axis intersects the surface at point *M*, and if *MN* is the normal to the paraboloid at *M*, the straight line that is symmetrical with *DM* in relation to *MN* to plane *DMN* passes through a fixed point *F* on the axis of the paraboloid. This point is called the focus (Fig. 45).

If the surface of the paraboloid has been covered with a substance which is such that the energy that passes through it by refraction is zero, all the energy of ray *D*, which we now assume to be a light ray, will be reflected along ray *MF*.

If a star is very distant, the light energy that we receive from it is slight. However, a mirror like the one that has just been described, if it has a large diameter, that is, a large enough sur-

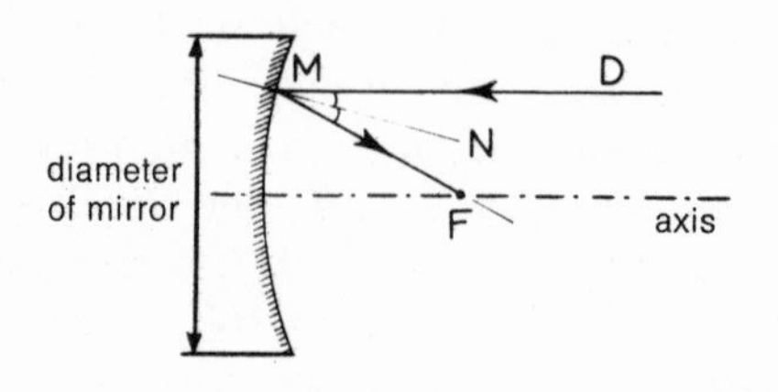

FIGURE 45

face, will receive sufficient energy from the star in the form of parallel light rays (the star is far enough away to be regarded as at an infinite distance) so that the star can be observed at focus *F*, either directly by the eye with a special lens (ocular), or by leaving an image on a photographic plate. This is the principle of the great modern telescopes, in which the diameter of the objective (that is, the mirror) may be several meters. It can be proved that the clarity of the instrument, that is, the ratio of flux emitted to flux received is, in the case of objects whose diameter is nonperceptible, like the stars, proportional to the surface of the objective.

b4. REFRACTING AND REFLECTING TELESCOPES. The term "lens" refers to a transparent cylinder, generally made of glass, bounded at each end by a portion of a sphere; the radii of the spherical portions of each end may be different (Fig. 46).

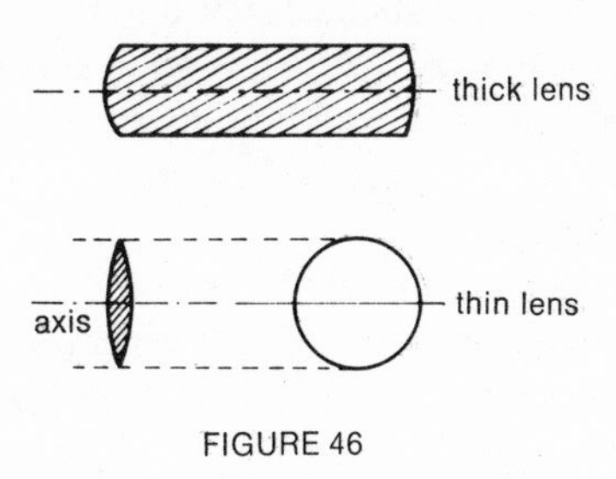

FIGURE 46

Generally speaking, it is more common to use thin lenses, that is, lenses in which the height of the cylinder (the thickness of the lens) is negligible in relation to the radii of curvature of the spherical portions. Thin lenses have optical properties which cause them to be classified in two categories: converging lenses and diverging lenses.

If we send a pencil of rays that are parallel to each other onto one of the faces of the lens, these rays, after undergoing refraction on each side of the lens, converge toward a point *f* as they emerge (Fig. 47). All points *f* which correspond to all the possible pencils of light rays are in the same plane which is parallel to the plane of the lens and is called the image focal plane.

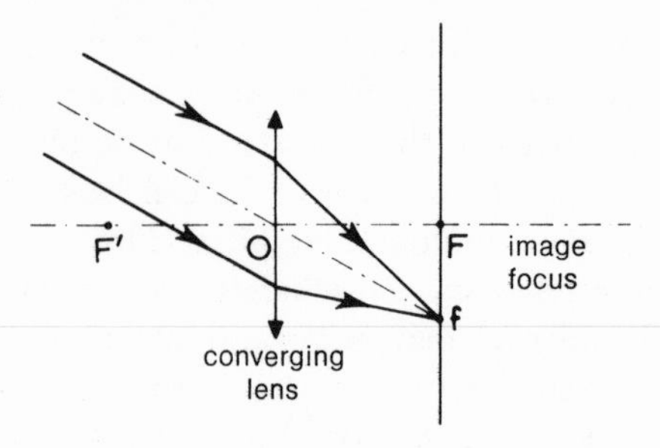

FIGURE 47

Specifically, point F located on the axis at the place where rays parallel to the axis converge after passing through the lens is called the image focus. If light rays all leave from the same point of a plane symmetrical with the focal image plane in relation to the lens, they will be parallel as they emerge (Fig. 48).

The new plane thus defined is the object focal plane, also known as the object focus. Diverging lenses are made so that a pencil of parallel rays diverge as they emerge as if they came from a point f (Fig. 49).

Let us consider a converging lens and a very remote object seen from the center O of the lens at a very small angle α. In the image focal plane of the lens there will be formed an image AF of the object which one will be able to observe at close range and consequently at a much larger angle than α (Fig. 50).

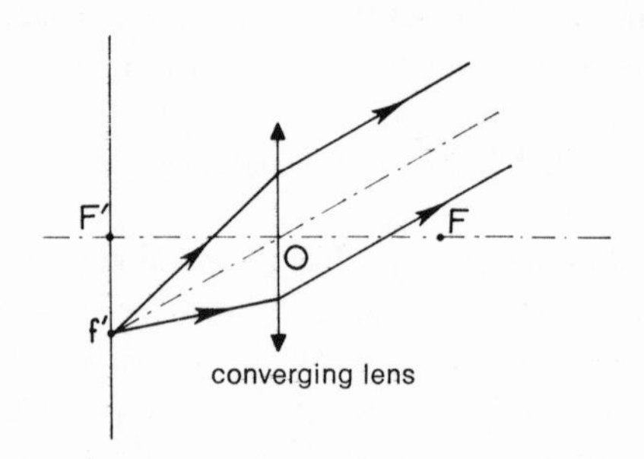

FIGURE 48

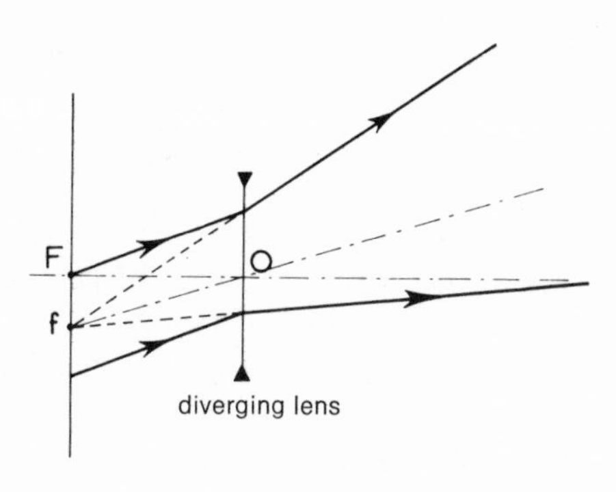

FIGURE 49

To do this, one will utilize a group of lenses forming a magnifying glass and known as an ocular. In this way we have constructed a refracting telescope. Instead of observing with the naked eye through the ocular, one can photograph the object by placing a photographic plate in the focal plane. The distance from the center O of a lens to the focus F is called the focal length of the lens. If we let f stand for this distance, the ocular will have a focal length of f'. It can be proved that the magnification M of the telescope, that is, the ratio between the angle at which one sees the object in the telescope and angle α, is

$$M = \frac{f}{f'}$$

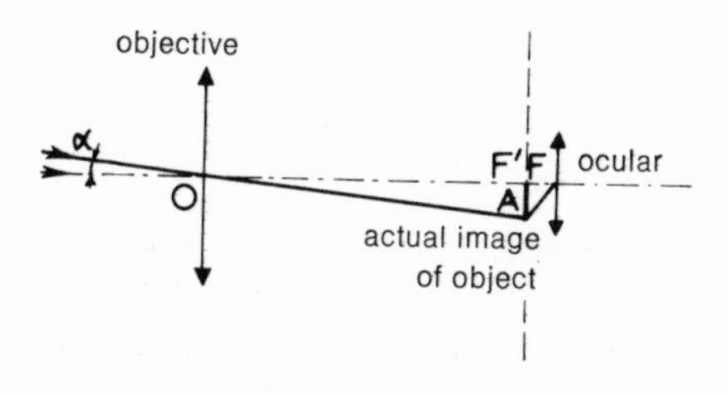

FIGURE 50

provided the object focus of the ocular has been made to coincide with the image focus of the objective.

If the object subtends a negligible angle α, but if it has a certain brilliance, its image at the focal plane will be a luminous point, and here again, the clarity of the instrument will be proportional to the square of the diameter of the objective, which makes it desirable to construct telescopes whose objectives have a large diameter. However, very large lenses are difficult to make. Moreover, since the index of refraction of glass varies with the wavelength of the light, the image contains iridescences caused by the superimposition of different images formed from the different colors of which the light coming from the object is composed. We say that the lens introduces chromatic aberrations.

For this reason it is preferable to utilize the parabolic mirror as an objective; now we have a reflecting telescope. The reflecting telescope is free of chromatic aberrations since the image is formed in the focal plane by reflection rather than by refraction.

To conclude these general remarks on astronomical instruments, we will say a word about resolving power. Because of the phenomenon of diffraction, the image of a point of light is not a point but a small circular area that is almost a dot surrounded by concentric circles that are alternately dark and light. If two objects in the sky are separated by a very small angle, the two images, since they do not appear as two distinct points, will be more or less merged and the observer will not be able to tell them apart. The term "resolving power" of an instrument refers to the smallest angle dividing two pointlike luminous objects that can be distinguished. It can be proved that this angle a is given by the formula

$$a = \frac{14}{D}$$

In this formula, a is expressed in seconds of arc and D is the diameter of the objective in centimeters. Thus, a telescope whose objective is 14 centimeters in diameter has a resolving power of one second of arc, which enables one to easily see the planet Uranus in the form of a disc, since its apparent diameter is on the order of three or four seconds of arc. Similarly, such a telescope will enable one to distinguish the elements of a system of double stars seen from the earth at an angle of one second.

Finally, here is a table illustrating the way in which the clarity of an instrument increases with the diameter of the objective. In order to understand this table, one should refer to the definition of the magnitude of a star given a little further on. For the moment, it is enough to know that the magnitude of a star characterizes its luminosity; the greater the magnitude, the dimmer the star. The most brilliant stars have magnitudes on the order of 0 or 1; the dimmest stars visible to the naked eye have magnitudes equal to about 6.5.

Here is a list of the magnitudes of the dimmest stars that can be observed with the instrument, in terms of the diameter of the objective:

DIAMETER OF THE OBJECTIVE	LIMITING MAGNITUDE
NAKED EYE	6.5
5 cm	10.5
10 cm	12.1
20 cm	13.6
50 cm	15.6
100 cm	17.1
200 cm	18.6
300 cm	19.6
500 cm	20.5

c. PHOTOMETRY

c1. LUMINOSITY OF A SOURCE. Given a point source of light S, this source is, as we have seen, the locus of a vibratory phenomenon whose amplitude is a and whose period is T. We shall agree that the energy of the source is proportional to the square of amplitude a. There is associated with the source a quantity I called the intensity of the source, which depends on the energy emitted by the source, and therefore on the square of a.

If now we receive on a screen (Fig. 51) the energy of a

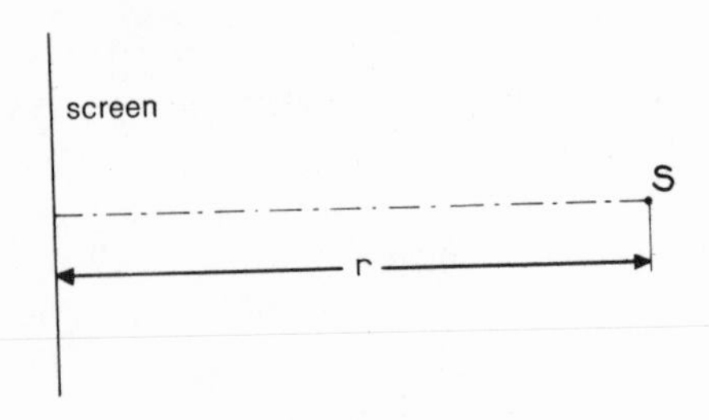

FIGURE 51

source of intensity I situated at distance r, this energy distributed over the surface of the screen produces what is called the illumination of the screen.

This measurable physical quantity—let us call it L—is related to I by the formula

$$L = \frac{I}{r^2}$$

L is also called the luminosity of source S.

c2. APPARENT MAGNITUDES OF THE STARS. Let us suppose that we are observing two sources S_1 and S_2; for example, two stars whose luminosities are L_1 and L_2 respectively. With each of these sources we associate a magnitude, m_1 for S_1 and m_2 for S_2, which is such that

$$m_1 - m_2 = -2.5 \log \frac{L_1}{L_2}$$

This relationship, which is known as Pogson's Law, defines magnitudes only in terms of their difference. One arbitrarily establishes the magnitude of a certain star by saying that if the luminosity of this star is L_2, its magnitude will be zero. Therefore, the magnitude m of a given star whose luminosity is L, in relation to L_2, will be

$$m = -2.5 \log L$$

It should be noted that the greater the luminosity L, the lesser the magnitude. The magnitude of the brightest star visible to the naked eye, Sirius, is -1.6; the dimmest stars visible to the naked eye have magnitudes in the neighborhood of $+6$.

3. ABSOLUTE MAGNITUDES. A star of luminosity L is located at distance r from the observer. Its intensity is

$$I = Lr^2$$

The quantity r will be measured in a unit called a parsec, which is defined as the distance of a star from which the orbit of the earth would be seen at an angle of one second of arc.

Now let us suppose that a star whose intensity is I is brought to a distance of 10 parsecs, its luminosity would then be L', and obviously

$$I = L'(10)^2$$

Its magnitude would then be M and we would have

$$m - M = -2.5 \log \frac{L}{L'}$$

or

$$m - M = -2.5 \log \frac{100}{r^2} = -5 \log \frac{10}{r} = 5 \log r - 5$$

since $\log 10 = 1$. The quantity M is known as the absolute magnitude of the star, and the difference $m - M$ is known as the *distance modulus*. If one has a way of finding M and of measuring m, one can then calculate the distance r of a star.

V. ASTRONOMY

a. DESCRIPTION OF THE SOLAR SYSTEM

a1. THE PLANETS. The description of the movements of the

planets on the celestial sphere given in the first part of this book was very much complicated by the fact that, in the first place, these movements were related to the center of the earth, and in the second place, we tried to represent them only in terms of circular and uniform motion.

The ideas of Copernicus and Kepler have enabled us to form a conception of the movements of the planets which is at once much simpler and much more rigorous.

All appearances are perfectly explained if we agree to take the center of the sun as the origin of the coordinates. On this basis Kepler was able to formulate his three famous laws: (1) The planets describe ellipses, one focus of which is occupied by the center of the sun. (2) As the planet moves, the radius joining the center of the sun to the center of the planet sweeps out equal areas in equal amounts of time. (3) The ratio of the cube of the semi-major axis to the square of the period is the same for all planets.

In order to explain these laws, we shall first give a few necessary concepts about ellipses (Fig. 52).

An ellipse is a plane curve which may be defined as the locus of a group of points M the sum of whose distances from two fixed points, called foci, is constant. In Fig. 52, if S and S' are the foci, when point M describes an ellipse it will always be true that

$$MS' + MS = 2a$$

a being a constant. Point O, the midpoint of SS', is a center of

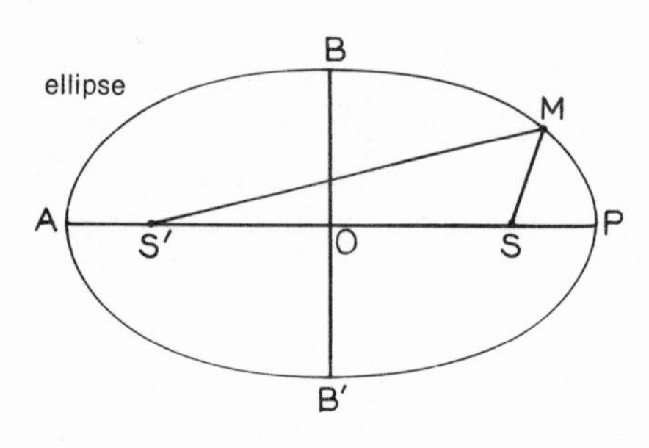

FIGURE 52

symmetry for the ellipse; similarly, line *S′S* and its perpendicular *B′B* drawn at *O* are axes of symmetry for the ellipse. Axis *S′S*, called the major axis, intersects the ellipse at two points, *A* and *P*. Kepler's first law states that a planet *M* describes an ellipse of which the sun occupies one focus; let this focus be *S*.

Point *P*, the point on the major axis closest to *S*, is called the perihelion; point *A* is the aphelion. Of course, since *P* is on the ellipse,

$$PS' + PS = 2a$$

Hence

$$OP = a$$

The quantity *a* is called the semi-major axis of the ellipse. The ratio, obviously smaller than one,

$$e = \frac{OS}{OP}$$

is the "eccentricity" of the ellipse. If *S′* and *S* merge, *O* merges with them and the ellipse becomes a locus of points *M* which is such that

$$OM = a$$

which means that the ellipse is now a circle whose center is *O*. *OS* is zero and so is eccentricity. The closer its eccentricity is to one, the flatter the ellipse; at the furthest extreme, as *e* tends toward one, the ellipse tends to merge with line segment *AP*.

Figure 53 shows three ellipses with the same semi-major axis *a* and with eccentricities of 0, 0.5, and 1.

Now let us see what Kepler's second law means.

Figure 54 shows an ellipse one of whose foci is occupied by the sun and which is described by a planet *M*. When *M* passes from point M_1 to point M_2, points rather close to the perihelion, the radius vector joining *S* to *M* passes through the crosshatched

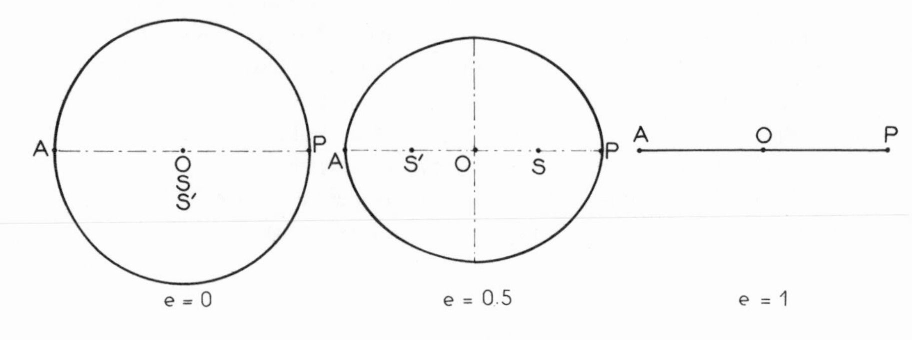

FIGURE 53

area SM_1M_2; when M goes from M'_1 to M'_2, the radius vector passes through the crosshatched area $SM'_1M'_2$. Kepler's second law states that if areas SM_1M_2 and $SM'_1M'_2$ are equal, the lengths of time it takes M to go from M_1 to M_2 and from M'_1 to M'_2 are equal. It follows immediately (Fig. 54) that the speed of point M on the orbit is greater between M_1 and M_2 than it is between M'_1 and M'_2. This law is known as the Law of Areas.

Finally, let T be the time required by planet M to travel around the ellipse; Kepler's third law states that

$$\frac{T^2}{a^3}$$

has the same value for all planets.

For example, for the earth, let us say that $a = 1$ (a is an astronomical unit equal to about 150,000,000 kilometers) and that

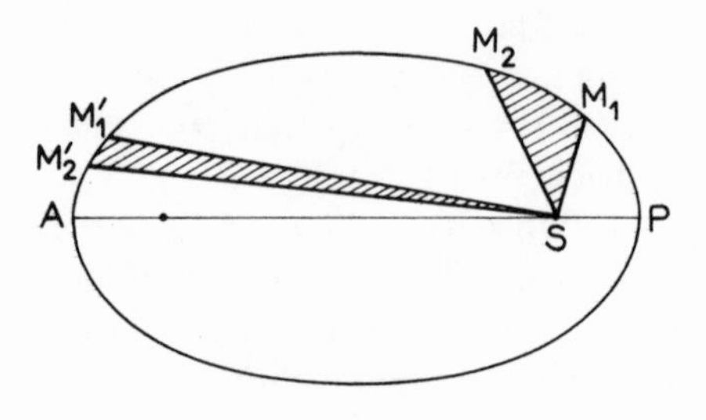

FIGURE 54

$T = 1$ (T is the sidereal year, which is equal to a little over 365.25 days). Now let us suppose that a planet M' has a sidereal revolution T' of 11.862 sidereal years, we will have:

$$\frac{T'^2}{a'^3} = \frac{T^2}{a^3} = 1$$

and therefore

$$a' = 5.2 \text{ astronomical units}$$

The solar system contains nine planets which are, in increasing order of their distances from the sun (or, more precisely, in increasing order of the semi-major axes of their orbits): Mercury, Venus, Earth, Mars, Jupiter, Saturn, Uranus, Neptune, and Pluto. To this list must be added several thousand small planets, of which the first four (Ceres, Juno, Pallas, and Vesta), which are the largest, although their radii are only on the order of a few hundred kilometers, were not discovered until the beginning of the nineteenth century. The orbits of these small planets are located between the orbit of Mars and that of Jupiter.

The planes of the orbits of the planets are only slightly inclined in relation to the plane of the earth's orbit, as one can see by consulting the table below. This table also tells us that the eccentricities of the planets are, generally speaking, very slight.

	a (IN ASTRONOMICAL UNITS)	e	i (INCLINATION OF ORBIT TO THE ECLIPTIC)
Mercury	0.3871	0.206	7°0′
Venus	0.7233	0.007	3°24′
Earth	1	0.017	0°
Mars	1.524	0.093	1°51′
Jupiter	5.203	0.048	1°19′
Saturn	9.555	0.056	2°30′
Uranus	19.22	0.046	0°46′
Neptune	30.11	0.009	1°47′
Pluto	39.49	0.250	17°10′

The semi-major axes of the orbits of the planets do not seem to be random quantities. The "Law of Titius-Bode" assigns them the following values:

$a_1 = 0.4$ (a_1 is the semi-major axis of the orbit of Mercury)

$a_n = 0.4 + 0.3 \times 2^{n-2}$ (n being greater than or equal to 2)

In the second formula, n is the position of the planet working outward from the sun, and a_n is the semi-major axis of its orbit expressed in astronomical units.

Here are the semi-major axes of the planets as they are given by "Bode's Law." We shall compare them with the values given above.

Mercury: $a_1 = 0.4$; Venus: $a_2 = 0.7$; Earth: $a_3 = 1$; Mars: $a_4 = 1.6$; the small planets: $a_5 = 2.8$; Jupiter: $a_6 = 5.2$; Saturn: $a_7 = 10$; Uranus: $a_8 = 19.6$; Neptune: $a_9 = 38.8$.

We see that these numbers are rather close to the values that were precisely determined, except in the case of Neptune. In the case of Pluto, we have not given the value provided by "Bode's Law," because it would be totally incorrect.

The value a_5 assigned to the small planets calls for an explanation. Obviously, not all the small planets have a semi-major axis of 2.8 astronomical units, but most of the values of their semi-major axes are grouped around this value. It is curious to note that "Bode's Law" was known long before the discovery of the small planets (the first one was seen by Giuseppe Piazzi on January 1, 1801). It was necessary to have the value 2.8 in the sequence for the law to be correct.

a2. THE SATELLITES. Mercury and Venus have no satellites. The earth, aside from the numerous artificial satellites that gravitate around it, has only one natural satellite, the moon. The moon has very significant dimensions for a satellite of the earth, for its apparent diameter is $\frac{3}{11}$ of the diameter of the earth. Its average distance from the earth is approximately 384,000 kilometers, but its elliptical orbit, one focus of which is the center of the earth, is considerably and rapidly distorted as a result of solar disturbances. We have described most of the irregularities of the lunar orbit (shifting of the node, of perigee, etc.) in Part I.

Mars has two satellites which are very small and very close

to the planet, Phobos and Deimos. Their orbits are almost circular and are in the same plane.

Jupiter has twelve satellites. The first four, numbered I to IV, are called Io, Europa, Ganymede, and Callisto. They are known as Galilean satellites because they were discovered by Galileo, in 1610. They can easily be seen with a good pair of binoculars. Their orbits are almost circular and more or less in the same plane, but since the masses of these satellites are not very small, they are subject to significant mutual perturbations which makes the study of their orbits difficult.

The other satellites of Jupiter are numbered V to XII, in order of their discovery. They are small and their orbits exhibit considerable variety. Satellites VIII, IX, XI, and XII are retrograde, whereas the rule in the solar system is that almost all bodies follow a direct motion around the principal body. The orbits of satellites VI and VII are very eccentric and very much inclined to the plane of the equator of Jupiter.

Saturn has the distinguishing characteristic of having rings composed of very small bodies not joined together, with the result that the outer edge of the ring turns less quickly than the inner edge. Saturn also has ten satellites. One of them, Phoebe, is retrograde; the largest is Titan.

Uranus has five satellites: Ariel, Umbriel, Titania, Oberon, and Miranda. The first four have orbits which are almost circular, retrograde, and considerably tilted in relation to the plane of the ecliptic.

Finally, Neptune has two satellites, Triton and Nereid. Triton has a retrograde motion; its orbit is almost circular. On the other hand, the orbit of Nereid has a high eccentricity.

a3. THE COMETS. Comets, which are often very spectacular, are bodies which seem to be very different from the planets. A comet consists of a nucleus and a coma, which form the head, and a tail, often very long, which gives the comet its characteristic appearance. The tail is extremely thin and the solar wind, carrying a magnetic field, drives back this tail, which is always opposite the sun. The nucleus is very small (a few kilometers in diameter at the maximum), but the coma is very wide and its diameter may reach 250,000 kilometers. It is nevertheless true that the mass of a comet is very small, altogether negligible in comparison with the masses of the planets and the larger satellites.

Their orbits are also very unusual. They are all very elongated, their eccentricities being very close to one. In other words, most of their orbits are quasi-parabolic. However, there is nothing supporting the statement that certain comets come from outside the solar system. If this were true, we would observe orbits that were clearly hyperbolic, which is not the case. If many comets whose orbits have been calculated during one of their appearances have not been seen again, this is probably due to the fact that the orbit was profoundly altered by the comet's passing in the vicinity of Jupiter and that identification thus becomes practically impossible.

b. THE ASTRONOMICAL UNIT

Defining the space in which the stars of the solar system swim as a Euclidean space of three dimensions enables one to carry out measurements in this space, particularly measurements of distance based on measurement of angles and of the lengths of common bases. Thus by sighting the moon from two points on the earth, one can find the distance of the moon. For more distant stars the problem is more difficult, for since a terrestrial base must necessarily be short, the straight lines drawn from the ends of the base to the star are almost parallel. However, mechanics suggests a system of fundamental units which is particularly effective in the solar system.

Kepler's third law is actually written

$$\frac{T^2}{a^3} = k(M + m)$$

T is the period of the planet, a the semi-major axis of its orbit, k the constant of universal gravitation, M the mass of the sun, and m the mass of the planet. First let us assume that m is negligible in relation to M, which is true at first approximation, since m/M does not exceed 1/1,000.

This gives us

$$\frac{T^2}{a^3} = kM$$

Let us take as a unit of mass the mass of the sun, $M = 1$; as a unit of time, the day. If T is the period of revolution of the earth around the sun and if the semi-major axis of the earth's orbit is taken as a unit, k is determined.

If we now take another planet, we will have

$$\frac{T'^2}{a'^3} = kM$$

and we will need only measure the period of its sidereal revolution T' in order to find its semi-major axis a' in astronomical units. In fact, the astronomical unit is not defined exactly this way; it is more precise to say that this unit is equal to the semi-major axis of a planet of negligible mass, not subject to perturbation, and whose sidereal revolution is 365.2568983263 days.

The number of significant figures given seems very high, but it is true that such a high degree of precision may be reached. In fact, the accumulated data from observations of the planets of the solar system over dozens of centuries, however imprecise some of these may have been, make it possible to achieve this result.

Thanks to this system of astronomical units, a relative precision of 10^{-8} may be achieved in the matter of distances in the solar system. On the other hand, if one wants to know these distances in kilometers, the precision falls to 10^{-3} (or possibly 10^{-4}, now that we are carrying out measurements of distance by radar). In this case it is necessary to go back to the system of bases and to measure, for example, the angle at which one would see the earth from the center of the sun. These measurements, which are very delicate, contain numerous errors.

PART THREE

THE MODERN CONCEPTION OF THE UNIVERSE

Philosophical Account

I. THE SECOND COSMOLOGICAL REVOLUTION

Between 1900 and 1924 a second cosmological revolution occurred, and no one yet knows what the final outcome will be. The very word revolution implies that some analogy exists between this rebirth of cosmological thought and the one that occurred at the beginning of the classical age. However, it is only an analogy, whose meaning and limitations must immediately be defined with some care. In the first place, an inseparable part of the transformation in our conception of the universe was the radical shift that occurred in the foundations of physical science. Classical physics at the beginning of the twentieth century was an edifice so vast, so complex, and so solid that it is impossible to enumerate all the hypotheses derived from its foundations and superstructure. We shall confine ourselves to mentioning those that have a direct and close connection to the rebirth of cosmology. But in any case, as in the seventeenth century, the renovation of cosmology would be incomprehensible without the renovation of physics.

Second, as in the dawn of classical science, a complete change in our view of the universe taken as a whole took place at the beginning of the twentieth century. In the ancient cosmology, the universe was regarded as a concrete, self-contained system associated with certain idealized geometric forms: the sphere and the circle. During the classical age, as we have seen, this image was replaced by the idea of a spatio-temporal framework that was empty, homogeneous, and infinite, filled with bodies which interacted, but whose distribution was virtually unknown and in any case had no necessary relation to the framework in which the objects were distributed and in which events took place. In the twentieth century we again encounter something of the ancient image, for the way in which bodies are distributed in the universe is logically and mathematically connected to its spatio-temporal form. The universe is once again a physical and geometric whole, a structure in principle perfectly defined; once again we can include in the same description both container and content, which are henceforth bound by a necessary relationship. The result is that cosmology is again a science possessing a specific and well-defined object: the universe.

An upheaval in physics, a reversal in our perspective of the universe as a whole: these are the essential elements of the analogy between the two revolutions. They also mark its limitations. For while classical physics irrevocably destroyed almost all of ancient physics, modern physics retains the whole of its inheritance from classical physics. Not one of the theories of this physics is *false;* all retain their validity, subject to certain hypotheses and at a certain level of approximation. Entire sectors of modern technology, even at its most refined, depend on the laws of classical physics. To be sure, classical science also preserved some elements of ancient science, such as the description of celestial phenomena on the sphere. It retained geometry, whose uses it had extended; but it was primarily concerned with mathematics and astronomy, not with physics; with descriptive structures, rather than with causal and explicative structures. But contemporary physics borrows from classical physics not only a body of results, but a method, procedures for translating relations of causality into mathematical terms which one cannot imagine physics doing without. The laws of modern physics rest on mathematical relationships describing phenomena on a spatial and temporal scale that is extremely small, and prediction is made possible by the adding up of elementary phenomena. The second revolution may have restored cosmology, but it did not restore the prevalence of cosmology over physics.

Some similarity might also be observed in the reasons for which and the manner in which these sudden alterations in the cosmic vision took place. In both cases, two causes—advances in observation and a profound modification of theoretical structures—were at the root of the process of subversion. At the end of ancient science the accumulation of new observations that were more and more difficult to reconcile with geocentrism finally rendered this theory improbable. At the beginning of the twentieth century, the hypothesis that all the stars one could observe belong to the galaxy, however prudent it may once have been, had become improbable.

But whereas the ancient cosmology had presented itself as a system establishing claims to be definitive, the classical pseudocosmology offered itself as admittedly incomplete. Merely negative reasons—unsolved problems, unproved statements, risks of paradoxes—were not enough to generate the idea of overthrowing a cosmological system, when that was precisely what did not exist. Before even the idea of a new system could be conceivable, there had to be a clear-cut advance beyond what was known, accepted, and understood, on the level of observation as well as on that of theory. At the beginning of the seventeenth century, partisans of the ancient cosmology said to those who were looking for another, "Our theory is true; your ideas are false, and furthermore, blasphemous." In the twentieth century, supporters of classicism might rather have said to the inventors of the new cosmology, "You are proposing a system of the world, but what need do we have for such a system?"

And yet modern cosmology exists, and history shows us that it was theoretical physics—or rather, as it happened, that it was almost a single man, Albert Einstein, who took the first step and, breaking with what appeared to be the soundest principles of the formal cosmology, transformed the *idea* of the universe. But in order for this new idea to be understood and finally, after considerable grumbling, accepted, or at least tolerated, it was necessary that the telescopic *image* of the universe be also transformed beyond question. That this actually happened is certainly one of the greatest favors that fortune has ever bestowed upon genius.

However imperfect the analogies between the first and the second cosmological revolutions may be, they are well enough founded to prove that the history of cosmology clearly denies the image of the continuous progress of the science of nature in a rational framework determined once and for all, or even in accordance with certain governing ideas that emerged in the classical age. To be sure, what has been established in science is preserved and what has been preserved may serve as a criterion for recognizing what once *was* science and what was not, but only in retrospect. For one cannot infer from past progress that progress must always be made in the direction in which it has already been made, or even that a return to ideas that have been abandoned is out of the question. A return is not necessarily a regression. The very idea of the "universe" was denounced in the nineteenth century as an illusion (notably by philosophers for whom the idea of "humanity" was not), and yet it has recovered its place among scientific ideas.

Also, it would not be irrelevant to review the meaning and importance of the statement that, in spite of the scientific revolution of the twentieth century, classical physics nevertheless retains all its validity. For there is a kind of fortunate paradox in a modern science which is capable of being transformed from top to bottom without losing any of its established truths. Let us not try to see this as the result of some "dialectic"; what is happening here has nothing to do with those verbal jugglings by which Hegel or one of his imitators claim to breathe new life into what their predecessors said, or what their adversaries say, by integrating it into a more advanced truth to which they hold the key. When the question of progress is raised about theories of physics, what happens is at the same time much more difficult to bring about in practice and much easier to understand after the fact. It is essentially—or rather it becomes so when the work is done—a question of order of magnitude and order of approximation. In an equation, when an expression occurs whose numerical value is very small in relation to the value which experiment currently assigns to the variables, one can simply cancel it. In this way one obtains an equation which correctly describes the phenomena at a suitable order of magnitude. In this way many "modern" equations become "classical." For example, the expression v^2/c^2, in which c is the speed of light in a vacuum and v the speed of a moving object in relation to the system of reference chosen, appears in the fundamental equations of the theory of relativity (I a2); by regarding it as zero one obtains, generally speaking, the corresponding classical equations. When a space ship flies toward the moon, its speed in relation to earth is less than 20 kilometers per second, and v^2/c^2 is then on an order of 1/100,000,000; astronautics, at its present stage, can also rest on classical mechanics and get along without relativity. Similarly, the fundamental constant of microphysics, Planck's constant, h, is so small that many classical equations in which it ought in principle to appear but from which it is absent are valid for phenomena of dimensions much greater than those of atoms. But naturally this is not true for all orders of magnitude, or for all phenomena which enter into the field of observation. Some of these phenomena, although they are more or less directly accessible to us (that is, necessarily very "gross"), nevertheless bear the characteristic marks of microphysical objects, or involve very high speeds; in these cases, the classical equations must be replaced by modern equations which are more exact and *which are based on altogether different principles.* The success of classical physics rests on the fact that there is often a certain *numerical* continuity between its results and those of modern physics, but this numerical continuity conceals a *logical* discontinuity that is abruptness itself. Faced with proving the validity of the theory of general relativity, Einstein had first to demonstrate (as he did with complete success) that his new theory gave the same results as classical mechanics, on the order of magnitude of almost all possible observations, for it was not at all evident that with principles so different one could obtain results so close. After that, of course, it was necessary to prove (as Einstein also succeeded in doing) that the new theory provided a likely explanation for phenomena which were totally inexplicable by classical mechanics and some of which, indeed, had never been observed (Ic4).

Cosmology is by definition one of those areas in which interpretation of observations can be done properly only with the aid of physical theories which are in principle irreconcilable with the formal cosmology of the classical era.

This has to do, first, with the enormous scale of observable phenomena and on a deeper level, with the fact that we are dealing with a science whose object is the whole universe. For it is clear that in a case like this notions of approximation, although naturally not excluded, cannot be applied in such a simple manner. For example, one can say that the distribution of matter in the universe is "approximately" homogeneous, but not that the universe is "approximately" infinite or that it has been found to be "approximately" in a physical state totally different from the one that we see.

II. THE THEORETICAL FOUNDATIONS OF MODERN COSMOLOGY

Hence our first concern will be to review the principal breaking points which the advances in physics in the twentieth century created in the foundations of classical physics. This enterprise is not easy because although we know perfectly well how the upheaval began, we still do not know how, when, or where it will end.

However, from the point of view of cosmology, all stages of the revolution are not, at least for the moment, of equal importance. From this point of view it is the first stage, the theory of special relativity, and especially the second, the theory of general relativity, that are decisive. The roles of the third, quantum mechanics, and *a fortiori*, of the fourth, the theory of elementary particles, are still difficult to evaluate.

SPECIAL RELATIVITY, THE FIRST BLOW TO THE CLASSICAL FORMAL COSMOLOGY (Ia)

The theory of special relativity was not invented for a cosmological purpose. Einstein's motivation was classical in that he was not trying to create or improve a theory of the universe but to give existing physical theories a simpler and firmer logical base, which would help to coordinate them better among themselves and also, of course (although the actual importance of this motive is difficult to appreciate), to account for certain experimental facts difficult to interpret "classically"—notably the famous experiment of Michelson and Morley (Ia1). The theory fully achieved this goal, but it did so only at the price of abandoning the Newtonian postulates on space and time which were absolutely fundamental to classical physics and pseudo-cosmology. So, it was with the theory of special relativity that the cosmological revolution logically began (although historically it did not begin until after the theory of general relativity).

To examine this matter more closely, we must therefore explain how and why this abandonment of the Newtonian principles occurred, how far it went, and by what these principles were replaced. In so doing we shall see (*a*) in what respects cosmology is potentially affected by this shift, and (*b*) why, logically and historically, its real influence on the theory of the universe was not and could not have been appreciated until after the second stage, that of general relativity.

The theory of special relativity logically and mathematically develops the consequences that result from the combining of two principles which Einstein rightly regarded as fundamental, in the state in which he encountered physical theory:

▪ *The principle of the constancy of the speed of light.* In a vacuum, this speed is independent of the relative speeds of the source emitting the light and of the observer who measures it; it is a universal constant.

▪ *The principle of relativity.* If two systems of coordinates are in rectilinear and uniform relative motion, the laws governing changes in condition of the physical systems remain the same, no matter which system of coordinates these changes are referred to.* It follows that no experiment can reveal which of the two is "really" at rest.

The first principle, the speed of light, may be regarded as experimentally established—and it could already be proved experimentally at the time that Einstein published his first paper.

As for the second principle, relativity, the position of classical physics toward it was very unsatisfactory from a logical point of view. Relativity was regarded as valid for mechanical phenomena (indeed, it is an obvious consequence of the principle of inertia and the fundamental laws of dynamics), but not for electromagnetic phenomena. The latter were currently being related to an "ether" that was absolutely immobile in relation to Newton's absolute space, and relative to which Michelson's experiment had been made precisely for the purpose of measuring the speed of the earth. The conceptual operation that required genius was that of dissociating the principle of the constancy of the speed of light from the hypothesis of an immobile "ether" to which it appeared to be bound by an absolute logical necessity. For the astonishing thing about the whole matter is that by combining two principles, each of which is perfectly plausible, Einstein drew conclusions which appear to be strange and even paradoxical, when in fact they are not at all paradoxical and are no stranger than many other things that science has managed to establish beyond question.

We shall not play the game of presenting apparent paradoxes only to show that they do not exist. The only thing that matters here is that Einstein's two principles, taken together, are incompatible with Newton's principles of space and time. Indeed:

▪ Newton's "absolute, true, and mathematical space," the universal container of all objects, is a pure fiction. For let us imagine a system of coordinates A_1 as being absolutely at rest in this space and another system A_2 as being in a motion of uniform translation in relation to A_1. According to the principle of relativity,

* Albert Einstein, *Elektrodynamik bewegter Körper*. This paper was published in 1905 in the *Annalen der Physik*. For reasons of convenience, we have changed the order of the two principles and have slightly modified Einstein's wording.

this hypothesis makes no sense because relativity denies, *in principle*, that *any experiment* can ever determine whether it is A_1 or A_2 which is at rest.*

Absolute space is therefore, *in principle*, impossible to locate.

▪ The same is true of Newton's "absolute, true, and mathematical time," for reasons which are a little more complex but which also result from the combining of these two principles. It is, in principle, impossible to define and locate a clock that would mark, unlike any other, the time of the universe.

▪ The measurement of an interval of time, like the measurement of a spatial dimension, cannot be "invariant"; it necessarily depends on the observer who is making the measurement.

When we recall that classical cosmology was essentially formal and that Newton's principles of space and time played a fundamental role in this cosmology, we understand that with the formulation of the theory of special relativity and the verification of its excellent experimental references (Ia1, 4), the cosmological problem found itself implicitly stated in nonclassical terms.

And yet we must note that historically, it was not until twelve years later, after completing the second stage, that of general relativity, that Einstein explicitly undertook the reform of cosmology. In the area of cosmology, the fact is that special relativity did not by itself permit that forward step which would force rational thought to abandon, not its "dogmatic sleep," but rather, in this case, its skeptical repose. In other words, from this point of view, which we share, special relativity is primarily important as a first stage in the formulation of general relativity.†

In order to understand this better, let us examine more closely not what the theory of special relativity destroyed, but what it substituted for the principles it abandoned. We shall see that it did not abolish the idea of a spatial and temporal frame-

* Let us note that the situation is not at all the same if A_1 is rotating around A_2; here, in special relativity as in classical mechanics, one can perfectly well determine which of the two systems is "really" rotating.

† This position is entirely different from that of physicists for whom special relativity is a very important theory in its own right and not merely a stage in the development of the theory of general relativity.

work that was empty and infinite—in other words, that of a formal cosmology—and that it actually contributed nothing that was very new to the interpretation of astronomical facts.

Special relativity does away with absolute space and absolute time, but it replaces these concepts with another absolute: space-time (Ia3). This space-time has many of the properties of ordinary space, but it has an added dimension: its elements, its "points," are not points in the ordinary sense, but "point-events." Let us consider a physical object that is very small and very coherent, for example, a nucleus of hydrogen, a "proton." In many cases, classical physics represents this nucleus geometrically by a *point.* In space-time, however, this proton is not represented by a point but by a *line* (its "world lines"), for even if one regards it as at rest, it is still drawn out in time as long as it continues to exist. A "point" of space-time is, therefore, an instantaneous event at a given (ordinary) point, not *this proton,* but *the event:* the "instantaneous shock between this proton and a neutron," or another body of the same kind.

Why is this space-time absolute? Because, following the "world lines" that physical objects trace, one can construct between two "point-events" a spatio-temporal "dimension," the "world interval." This interval is a metrical invariant in the sense that its size is the same for all systems of coordinates belonging to a certain family, the family of the "inertial systems," each pair of which is in rectilinear and uniform relative translation. In terms of procedures for physical measurement, space-time changes nothing. There is no special instrument for measuring the world interval; its measurement is obtained by combining, by means of formulas characteristic of the theory, measurements of length and measurements of time. Thus it generally happens that two observers bound to different systems of inertia will be in disagreement on their separate measurements of length and time, but they will find themselves in agreement once they have calculated, on the basis of their measurements, the world interval between any two point-events.

Now let us suppose that we have identified *one* inertial system in the universe;* all physical systems which are in uniform

* This identification raises no practical problem when it comes to applying

translation in relation to this one will also be inertial systems, their descriptions of the universe will be equivalent in space-time, and they will be represented by a family of *straight lines.** This group of physical systems, this family of straight lines of space-time, plays the same role in the physics of special relativity that absolute space and time play in Newton's theory.

What are the implicit consequences of this state of affairs for cosmology? Cosmology has lost two absolutes, space and time; it has regained one, space-time. It is a definitive acquisition; indeed, on this point, general relativity will not change the situation: cosmology will remain, or rather will become, a theory of space-time. But from the standpoint of general relativity, space-time as special relativity defines it is only *one* of the possibilities, which we shall call Minkowski's space-time† (Ia3) and which will have little chance of being retained, for on this point special relativity remains classical. Like Newton's space and time, Minkowski's space-time—whose metrical properties are, moreover, similar to those of Euclidean space—is a geometric form independent of the material content of the universe. If we confine ourselves to special relativity, there is nothing that would have forced cosmology to emerge from its formal classical perspective.

On the practical level, moreover, the level of physical cosmology—that is, the description of the universe on the astronomic scale—special relativity invited no agonizing reappraisal. All the cosmic references that classical astronomy chose as concrete landmarks for absolute space are valid as inertial systems in special relativity. Moreover, since the velocities of celestial bodies are very low in comparison with the speed of light,‡ all the classical laws worked for the practical purposes of astronomers before the discoveries that inaugurated the new era in the cosmology of observation.

---

special relativity to a physical problem; from the point of view of cosmology, however, it raises an interesting theoretical one.

* This family does not contain *all* the straight lines in space-time, for certain lines in space-time cannot represent any object or physical process (Ia3, Fig. 57).

† After the mathematician who first (in 1908, three years after Einstein's first dissertation) gave this geometric interpretation of special relativity.

‡ In microphysics, however, velocities have been observed which are high enough to reveal the effects characteristic of special relativity (Ia3).

GENERAL RELATIVITY; DISCOVERY OF THE RELATIONSHIP BETWEEN MATERIAL CONTENT AND SPATIO-TEMPORAL FORM (Ib–c)

It was not with special relativity, but with the more advanced form of the theory that is traditionally called general relativity—a term which is being replaced by the "relativistic theory of gravitation"*—that the way opened for the new cosmology.

What is at issue here is not only the coordination of the measurement of space and time and hence the mathematical properties of the metrical framework of the universe; it is the relationship, which is both mathematical and causal, between the spatio-temporal form and its material content. This is an entirely new question and one without any classical equivalent, since in classical physics, and even in special relativity, as we have stated, the presence of the content has no effect on the properties of the container; it is, moreover, a question that is decisive for cosmology. The universe, instead of being a more or less indeterminate object located in a framework defined *a priori*, becomes a spatial and temporal structure whose metrical properties cannot be exactly known unless the properties of the distribution of the matter it contains are also known.

Here too, however, the development was not originally motivated by the felt need for a new cosmology. Here, too, it was a question of correcting certain flaws that had in fact long been known, in the foundations of classical physics and also in special relativity, which made the problems more acute. But this time the correction was to rock formal cosmology so profoundly that it made the classical vision of the universe unacceptable.

The whole of Galileo's mechanics rests on an intuition for

* This distinction is made for historical reasons which we should mention briefly because they are not without importance for cosmology. When he formulated the equations of the theory of general relativity, Einstein thought they must be capable of being completed in a way that would make them apply not only to the forces of gravitation, but to all physical forces (for example, electromagnetic forces). But on this point his hopes have been disappointed; although his equations describe the effects of gravitation perfectly, they are not adequate to describe the effects of the other physical forces.

which he could not provide the definitive justification, but whose profound rightness is in fact proved by relativistic mechanics: the intuition that the question of falling bodies is essential for anyone who is trying to establish the foundations of the science of motion. However, the principles of the classical science of Newton led to a clear distinction between the two questions (laws of the science of motion, on the one hand, and the laws of falling bodies on the other), for classical science is, in principle, independent of the existence and, *a fortiori*, of the laws of gravity. In this science the law of gravitation, of which the case of falling bodies is a particular instance, determines the effects of a certain set of forces whose origin is unknown and whose mode of action exhibits no necessity that is common to the general laws of motion.

But this conception of gravitation, which permitted the rise of classical mechanics and assured its extraordinary success, left a residue of irrationality which, after being tolerated for two centuries, had ceased to be acceptable to those who, as Einstein said of himself, were capable of "sniffing out and tracking down the ideas that led to the bottoms of things."*

Not that Einstein was searching, as so many others had done without success, for a mechanical explanation of gravitation; on the contrary, he openly turned his back on mechanics. But what seemed to him irrational in the classical edifice was the fact that it reserved separate places and in some sense different treatment for two phenomena which in fact are always associated: *inertia* and *gravitation.* To be sure, inertia is a universal property of bodies, all of which resist all efforts to set them in motion if they are at rest and to alter their motion if they are in motion. But gravitation is no less universal, no body escapes it, there is no screen by which one can protect oneself from it, and one can rid oneself of it momentarily and locally only by invoking other forces, notably, the forces of inertia.

Nevertheless, and for excellent reasons, classical mechanics reserved very different treatment for inertia and for gravitation. In Newton's theory, inertia is an intrinsic property of each body,

* "Autobiographical Notes," in *Einstein, Philosopher-Scientist* (New York: Tudor, 1949), page 16.

which possesses a certain invariable quantity of it (its "inertial mass") quite independently of other bodies. Whether a body is alone in the universe or whether there are thousands of millions of others in no way affects its inertia. When it is subject only to its own inertia, a body is at rest or moves in a straight line, by uniform motion; in relation to what? In relation to absolute space in the classical theory; in relation to any other body, which is itself reduced to its inertia, in the theory of special relativity.

Gravitation, on the other hand, is a force of *interaction;* in a world in which there existed only one body, gravitation would not exist, any more than love or hate would exist in a society that had been reduced to a single individual. Besides, this interaction depends not only, in the case of two bodies, on an intrinsic property of each body (its gravitational mass), but on the distance between them. In Newton's *Principia,* the law of inertia is present from the beginning, as an *axiom.* The law of gravitation, however, appears much later, as a *hypothesis* which provides a suitable explanation for the movements of the moon and the planets. But is this separation logically defensible? Whether it is or not, it leads to some very curious consequences. In the first place, Newton had good reasons for suspecting—and physicists were later to demonstrate experimentally—that the *inertial mass* of a body (which measures its inertia) and its *gravitational mass* (which measures its response to the forces of gravitation) are absolutely identical. Indeed, this follows from Galileo's hypothesis that in a vacuum, all bodies fall at the same speed. Is this not a serious indication that the two phenomena are fundamentally of the same nature?

Furthermore, experiments show that the forces of inertia combine with the forces of gravitation in exactly the same way as any other force. The fact has now been verified: a passenger in an artificial satellite of the earth is liberated from gravity (that is, from the force of gravitation exerted by the earth) not because gravity has ceased to exist where he now is, but because at every moment it is exactly counterbalanced by the force of inertia resulting from the motion of the satellite.

Moreover, the way in which the principle of inertia is formulated contains a kind of inner inconsistency. In order for a body to follow its path of inertia, a straight line traveled by uniform motion, it must be infinitely distant from any other body, otherwise its motion would be altered by the gravitational force exercised by that body. But if it is infinitely distant from any other body, how can we tell that its path is a straight line, since we possess no observable reference?

These somewhat abstract arguments made very little impression on the theorists and practicians of classical mechanics, but once one has accepted the principles and consequences of the theory of special relativity, they take on an importance that is altogether different. In effect, if this theory seemed to have no direct consequences for the question of gravitation, it had, on the other hand, some very immediate effects on the theory of inertia. Indeed, it follows from the principles of the theory that the inertial mass of a body ceases to be a constant, and hence objective, property of that body. In fact, it depends on the movement of this body in relation to the observer who is making the measurement. For example, when a physicist inserts particles into an accelerator in order to bombard atomic nuclei, the inertia of the particles increases in proportion to their velocity. In order to gain an additional degree of velocity, he must pay in terms of inertia at a higher and higher rate; but of course, for an observer who was traveling along with the particle, its mass would remain constant; this would be the "rest mass" whose value is shown in the tables.

Since inertial mass was now no longer constant, was it necessary for gravitational mass to be constant or not? In either case, the law of gravitation had to be revised, or rather "rethought." This gives us at least a vague idea of why Einstein, after completing the theory of special relativity, undertook a total restructuring of mechanics—an enterprise which after the fact seems characterized by an audacity and a difficulty that are truly incredible. In so doing he relied on the *principle of equivalence* between the forces of inertia and the forces of gravitation, a principle which is itself closely associated with the generalization of the principle of relativity; once inertia has lost its unique position with respect to gravitation, the family of fundamental observations bound to the systems of inertia ought, in principle, to lose all their privileges.

This variability of the inertial mass is, moreover, only one aspect of a much more general consequence of the principles of special relativity. Mass itself, the most tangible property of bodies, is merely a form of that less substantial property of physical objects, *energy*, a quantity which is by definition indeterminate "in itself," since it can only be measured by its transformations (Ia4). Inertia, according to the theory of special relativity, is no longer the exclusive property of matter "condensed" into a mass, that big stone which Dr. Johnson kicked in a rage to refute Berkeley's idealism, which denied the reality of matter. All energy, light for example, is equivalent to a mass, and the sun illuminates us by consuming a little of its enormous mass (Id2–4), but it takes very little mass to produce a large amount of radiant energy.*

But why and how is this of concern for cosmology? For two reasons which will now become clear:

▪ Gravitation is—along with inertia, of course—the only force that governs the movements of bodies at very great distances. Electric and magnetic forces, which are much more intense but *polarized* (gravitation always attracts, electromagnetism may attract or repel), play a very important role in phenomena on a small scale—atomic or subatomic—but on a large scale, they are, generally speaking, neutralized. Gravitation plays no role in chemical combinations, which are principally governed by electrical forces. On the other hand, it is the gravitational force of the sun that produces the annual motion of the earth, whereas the electromagnetic phenomena of enormous intensity that occur on the surface of the sun have only limited effects on the earth† (the aurora borealis, for example).

Thus the whole theory of gravitation has implications for how we are to conceive of the distribution of matter in the uni-

* The inertia of energy is an idea so fantastic, according to the principles of classical physics, that Einstein published this consequence, which he arrived at by sound logic, only in an appendix to his first dissertation, and in the form of a question; but it has been brilliantly confirmed by all of modern physics.

† We should note, however, that astrophysicists now think that magnetic fields that are very weak but extended throughout immense volumes and acting over very long periods of time play an important role in the distribution and movements of the diffuse matter in the galaxies.

verse, and we have seen that in this respect the Newtonian theory did not lead to very satisfactory conclusions.

▪ But the principal reason for the cosmological implications of the theory of general relativity we already know; this is that the theory involves a profound alteration of traditional conceptions of geometry and chronometry, because it associates the metrical properties of the container with those of the content (Ic3). But why should a theory that begins with *dynamic* principles (the relation between the forces of inertia and the forces of gravitation) have *geometric* consequences? This would be rather difficult to explain without making reference to very detailed considerations based on the *ad hoc* mathematical apparatus. Let us confine ourselves to a few simple remarks which will give a general idea of the train of thought that connects dynamics to geometry, forces to metrics.

For classical mechanics and special relativity, the path of inertia of a body is a straight line, but the path of a body subject to gravitation, for example, the earth's, is not a straight line at all. Now, if we consider that inertia and gravitation, in fact inseparable, are essentially of the same nature, we cannot consider that the straight line has any particular role to play, and to tell the truth, there is no longer such a thing as a straight line at all, physically speaking, since all bodies are subject to gravitation. The straight line no longer has any physical reality except as a limiting case, the tangent, corresponding to the limiting case of pure inertia, corresponding also to a very short segment of a path that is very slightly curved. But there is a *law* of gravitation; the paths of bodies that gravitate must be determined by this law and must be distinguished from the paths that would be imposed on them by other forces of any kind. In other words, these paths must have special geometric properties, analogous to the properties of straight lines in ordinary geometry. In particular, they must be the lines that can be traveled in the least time between two points; other reasons lead to the conclusion that light rays are not straight lines, but curves of the same family.

It is in this way that one is led to abandon traditional Euclidean geometry as the most appropriate mathematical form for representing the metrical properties of physical space, and to substitute for it a more complex metrical geometry, the so-called Riemannian geometry.* But, whereas the metrical properties of Euclidean space are perfectly determined by the axioms (once one has accepted the postulate about parallel lines), in the geometry of Riemann, metrical relationships are only defined in terms of a group of arbitrary functions. There is an *infinity* of Riemannian metrical systems that are mathematically possible, and this corresponds on the physical level to the fact that the action of gravitation is not uniform, but depends on the distribution of masses in the vicinity of the body whose path is being studied (the "test body," as it is called). In other words, geometry reflects the condition of matter in a certain region. Geometry is "local," and if a universal geometry exists that underlies all the local geometries, it must reflect the condition of matter *on the scale of the universe.*

These, presented briefly and in outline form, are the leading ideas of the theory of general relativity which Einstein founded between 1911 and 1915 and which have seen considerable development since then. What we have here is no unfounded theory, developed on a provisional basis. Einstein completely succeeded, at least as far as gravitation is concerned, in reaching the goal he had set himself, namely:

▪ to show that, by starting with completely different principles, the new theory enables one to deduce, at first approximation, all the phenomena explained by classical mechanics.

▪ to show that, beyond a certain degree of precision, certain phenomena, of which some had already been observed, and others were not yet known at the time he constructed his theory, are predictable on the basis of this theory but not on the basis of the classical theory. These phenomena, although rare and very subtle, have now been well established by astronomical observation and physical experimentation (Ic4). We shall not insist on this point, but shall confine ourselves to noting that general relativity, or the relativistic theory of gravitation, presents itself, as matters now stand, with all guarantees of conceptual and mathematical coherence and experimental confirmation that can be demanded of

* To be accurate, one should say "space-time" instead of "space," and "geochronometry" instead of "geometry." What has been changed is not Euclidean geometry, but the quasi-Euclidean geochronometry of Minkowski; but this impropriety of language does not alter the essential ideas (Ib).

a scientific theory. The only criticism one can make of it is that contrary to Einstein's expectation, it remains limited to the description of the phenomena of gravitation, but from the point of view of cosmology this is still the essential thing, for the reasons we have given.

Let us now examine the consequences of this theory for cosmology. They are obviously of considerable importance, for it is no longer possible to conceive of a formal framework in the universe that is independent of the determination of the actual content of the universe in terms of matter and energy. In other words, formal cosmology has become impossible. The theory of general relativity rules out the classical idea of a universal metrical framework which is defined *a priori* and in which one can, in principle, place any group of objects and any sequence of events.

As a matter of fact, it is no longer even certain, *a priori*, that all parts of the universe can be made to agree geometrically in a regular way, for, as we have said, geometry is "local." The sun, a rather massive body, "creates" in its vicinity a geometry whose fundamental curves (the equivalent of the Euclidean straight lines) are the paths of the planets (which are *very close to* Kepler's ellipses) (Ic4). But elsewhere, in the vicinity of Sirius, for example, another geometry will prevail. What happens when a body traveling from the sun toward Sirius changes geometry? Although in this particular case, the classical expectations, which are much simpler, are valid in practice, the mathematical problem is by no means easy, and in certain cases strange phenomena may be foreseen. In the vicinity of, or in the interior of, a mass, space becomes more "curved" as the mass becomes greater and more dense. If its dimensions exceed a certain limit, the curvature becomes such that space somehow closes in on itself; nothing, neither matter nor light, can get out of it, it is a part of the universe totally separate from the rest.* This seems to belong to the realm of science fiction, but it is not out of the question—and

* Eddington gives an example of this which, although completely imaginary, is rather effective. He calculates that a globe of water with a radius of 570,000,000 kilometers would form a space completely closed in on itself. Nothing could either enter it or leave it; a light ray would turn in a circle, returning indefinitely to its point of departure (Eddington, *Space, Time, and Gravitation*, Cambridge: Cambridge University Press; 1920, Chapter IX).

it has been much discussed, these last few years—that the required conditions might be realized at the end of stellar processes which may not even be exceptional.

It makes little difference, however, for the cosmological questions raised by general relativity are much more general in scope. A plurality of geometries is not easily tolerated as soon as one wishes to rise to a conception of the universe in which the procedures for measurement and the laws of physics are not limited to one small compartment of the universe. Einstein, like Galileo, Descartes, Newton, and Laplace, like all the great masters of natural philosophy, was convinced that a physics worthy of the name must be capable of extending its laws to all phenomena that have been or ever will be accessible to human observation or experiment, from the finest particles of matter to the most remote and enormous celestial objects. But his own attempt, so amazingly successful, to reform classical physics led him to raise problems unknown to this physics: the problem of bringing local geometries into agreement with one another, the possible existence of a universal geometry with which they would all agree, the structure of this geometry, and its relation to the overall distribution of matter and energy.

General relativity is only the second phase in the revolution in physics that has occurred in the twentieth century. In many respects the third phase, the construction of the quantum theories of matter and energy, and the fourth phase, now in progress, the theory of elementary particles, may be regarded as the most important, and no one can foresee what their effects on cosmological thought will be and just how far these may lead. Up to the present, these effects remain limited and primarily indirect as they make themselves felt through astrophysics, although the situation is obviously in process of evolving in this respect. We shall not undertake the presentation of the leading ideas of quantum physics, always a delicate operation even when intentionally brief, nor, *a fortiori*, of the physics of elementary particles, although it is these ideas that provide the key to the understanding acquired in the last forty years of the structure, origin, and evolution of the stars, the development of atomic species from one another, and perhaps the coherent description of the physical phenomena characteristic of a completed state of the universe.

Historically and logically, then, it is to the principles and consequences of the theory of general relativity that we must refer (and indeed they are sufficient in the present state of our knowledge) in order to understand the origin and foundations of modern theoretical cosmology. But before discussing the results of this theory and the problems of this new cosmology, we must examine in what way and to what extent the advances in astronomical observation and in experimental physics have contributed to it.

III. THE MODERN IMAGE OF THE UNIVERSE (II)

THE TRANSFIGURATION OF THE APPEARANCE OF THE COSMOS

As we said earlier, by 1920 astronomy had acquired a very precise knowledge not only of the solar system but also of the galaxy, that enormous mass of stars whose existence and dimensions Galileo was the first to suspect, and whose systematic exploration Herschel had undertaken at the beginning of the nineteenth century. What we know about it now is much more extensive and precise, but not essentially different from what was known fifty years ago, and the important thing, if one wishes to understand the contribution of modern astronomical observation to cosmology, is to note carefully the two limits of knowledge at that time: (1) doubt about the question of whether one can observe something beyond the frontiers of the galaxy; and (2) ignorance—still very profound—about the physical nature of the stars, the sources of their radiance, their origin, and their evolution.

On these two points, the advances have been decisive for the renaissance of cosmology. On the first, the information acquired and consolidated during the past fifty years is almost impossible to interpret in classical terms, and strongly suggests a cosmology consistent with the principles of general relativity. On the second, the discovery of the secret of the stars is opening the way for a historical interpretation of cosmic events in which all physical systems, from atoms to groups of galaxies, are part of a total evolution which may be unique and irreversible.

As for the discovery of the world outside our galaxy, the decisive step was taken in 1924 when the American astronomer Edwin Hubble, thanks to the completion of a new telescope,* put an end to a hundred-year-old controversy by demonstrating beyond all possible argument that the great nebula of the constellation of Andromeda was a mass of stars comparable to our galaxy. It followed from this that the majority of observable nebulae were *galaxies*,† that the gaze of astronomy penetrated vastly beyond the frontiers of the galaxy, and that the latter, no longer identified with the universe, was becoming a "local" and probably a minor object in terms of its situation, dimensions, and physical characteristics.

Although this event belongs to a time that is already distant in terms of the present rate of scientific progress, one cannot really understand contemporary cosmology unless one has an accurate appreciation of the new situation and problems with which Hubble's discovery confronted astronomy. A large telescope reveals a teeming mass of galaxies: they are literally innumerable (IIc1). There are galaxies of all dimensions: some have large apparent diameters;‡ others can barely be distinguished from the image of a dim star. It was literally a new universe, the realm of the nebulae, to use Hubble's expression, that was opened up to exploration. But *how* was one to go about this exploration? Not only were there obviously very great technical difficulties to be overcome, but a problem of interpretation, an epistemological problem of formidable dimensions appeared, or reappeared. Out

* This was the Hooker telescope on Mount Wilson which, with its 100-inch mirror, was the largest in the world until 1948 and which, in 1975, is still third largest after the Hale telescope on Mount Palomar (200 inches) and the one on Mount Hamilton (120 inches). The three largest telescopes are located in the mountains of California, where the climate is particularly favorable for astronomical observation.

† The others are either very dense groups of stars, or masses of diffuse matter within the galaxy.

‡ For a telescope, of course. All galaxies are invisible to the naked eye, except the Magellanic Clouds, observable in the southern hemisphere, and the Andromeda nebula, which a sharp eye can distinguish on a cloudless and moonless night in our latitudes.

of this profusion of disparate and, for the most part, very dim images, how was one to derive with any certainty a picture of a physical system whose elements have known properties and whose actual distribution in space can be determined?

In spite of the creation of very large instruments, the situation of astronomy in the twentieth century remains essentially what it has always been. Space has three dimensions; time, one; but astronomy sees the content of extraterrestrial space-time only as it is projected on the two dimensions of the celestial sphere that it has inherited from the ancient cosmology. The vast and patient labor by which the physical structure first of the solar system and then of the galaxy had been reconstructed from celestial appearances had therefore to be taken up again and completely revised for the world of the galaxies, with the additional complication that the temporal dimension, which until then had been overlooked without much risk, had now to be taken into consideration.* It was an enormous undertaking in which one could not take a single step without some working hypotheses.

In fact, first appearances very soon set Hubble working in a direction from which—and this is an important fact—research has not to date had occasion to deviate substantially; namely, the direction indicated by the following hypotheses:

▪ The system of the galaxies is formed of elements that are similar enough qualitatively and quantitatively to be compared with one another in our representation of the system as a whole.

▪ These elements are distributed uniformly in space. Once this hypothesis is accepted it becomes possible to reason by analogy, by passing from the known to the unknown, from the near to the far, and to extend the analogy step by step. One takes advantage of our knowledge of the galaxy to interpret the images of the nearest galaxies, and then, with the aid of this interpreta-

* The time lag between emission and reception of light does not exceed a few hours for the most remote objects in the solar system. It is several times ten thousand years for the remote objects of the galaxy; this is still negligible in relation to the duration of the great astrophysical processes. But the lag greatly exceeds a billion years for the very remote galaxies, and as we shall see, it is no longer negligible at all in the history of the stars, and perhaps of the universe.

tion, one interprets the images of more remote galaxies. This knowledge includes, for example, knowledge of the dimensions of the galaxy, of its stellar content, of the intrinsic luminosity of certain stars, of the cycles of other stars which exhibit extremely regular variations in luminosity (IIa2). All these data enable us to estimate the stellar content, dimensions, distance, and overall intrinsic luminosity of the nearest galaxies. The most brilliant stars that are revealed there will serve as a standard by which to estimate the distance of somewhat more remote galaxies in which stars assumed to be of the same type are still visible. Next, it is the overall luminosity and the overall form of the galaxies that are considered, for it becomes impossible to distinguish their stars. Finally, since the number of galaxies observed increases simultaneously as their images become dimmer and more difficult to analyze, methods of enumeration enter into play. It is obvious that in working this way one risks falling into a vicious circle, for in order to advance in observation, it is necessary to introduce hypotheses which only new advances in observation could truly confirm.

For example, by making the hypothesis that the ten most brilliant stars of galaxy B, whose distance is unknown, have on average the same intrinsic luminosity as the ten most brilliant stars of galaxy A, which is nearer and whose distance is known, we obtain a means of calculating the distance of galaxy B. But to be able to judge whether and to what degree the hypothesis is correct, we would have to know the distance of B. Thus one runs the risk of discovering in the world of galaxies only what one has put into it.

But the science of nature is not logic; it is not so completely trapped in circles. The multiplicity of hypotheses and avenues of approach makes it possible to mitigate this disadvantage to a certain extent. Besides, the results of new observations are never exactly what one was expecting; they only rarely supply a yes or no answer to the questions one had in mind, but the accumulation of these results gradually confirms or, on the contrary, wears out or exhausts the hypotheses. If the working hypotheses one has introduced are totally inadequate, one will eventually become aware of this; if they are only partially inadequate, the results will enable one to correct them with a view to a later exploration which will in turn rectify them still further. An effect of convergence by successive approximations *may* occur and confer a certain authority on the assertion of a well-tested hypothesis. But of course the success of such a procedure is not guaranteed in advance, as it is in the case of a mathematical theory which makes use of the method because it can define *a priori* the conditions of convergence.

In this respect one can now say that after much trial and error, after long and lively controversies among astronomers, success has crowned the undertaking of Hubble* and his successors. *Except for one essential discovery that was altogether unexpected and altogether contrary to the original hypotheses*, to which we shall return a little later, Hubble's initial hypotheses have stood up well. To be sure, the regularity and uniformity that he had assumed in the kingdom of the nebulae are not as perfect as he believed them to be—in this respect his hypotheses are rather crude and partially inaccurate—but they are not essentially wrong and they have survived, with appropriate corrections. In spite of the accumulation of facts and the introduction of new and more powerful instruments, nothing to date has forced us to dismiss the approximate image—at that time an entirely new image—of the universe to which the 1924 discovery of the kingdom of the nebulae had immediately given rise. This is critical, for this discovery introduced into our representation of the world something just as important and irreversible as the revelations of Galileo's telescope.

The dominant fact is that, on a very large scale, the distribution of matter in the universe is uniform. Not only is it established that neither the earth, the sun, nor the galaxy is the center of the universe or occupies any remarkable position in it, but that the universe has no center, no remarkable position, no privileged place. If we place an observer, not, as Voltaire did, on Sirius, but in the vicinity of a star located in one of the countless remote

* Hubble, who had come late to astronomy, had numerous and eminent adversaries both at home and abroad who did not deal gently with him; which makes the final success of his enterprise all the more convincing.

galaxies seen by the telescope on Mount Palomar; if we assume that he has at his disposal astronomical instruments comparable to our own; perhaps what this observer will see in his vicinity will be unimaginably different from what surrounds us, but if he looks far enough into the sky, the world he will observe will be similar to the world that we can see since Hubble—and the further he looks, the more similar it will be.

This is the most striking feature of the universe, which year after year, for half a century, has repeatedly been confirmed; the diversity of things and events in the world is practically infinite, and one is not taking any great risk in saying so; if astronomers continue to perfect their instruments and, above all, if they are able to install their observatories beyond the atmosphere, they will have an enormous number of new and surprising things to discover; however, on a very large scale, the differences are blurred and the universe becomes a homogeneous and uniform system. And after all, there is nothing very extraordinary about this. Seen at close range, the surface of the earth, bristling with mountains, dotted with valleys and caverns, furrowed by rivers and glaciers, covered with different kinds of vegetation, is infinitely varied; seen from the moon, the earth is a very simple and very regular geometric figure, and so astronomers conceive of it. The difference is that the geometric form of the earth has been known for a long time and that there is little reason to expect that the progress of knowledge will alter the situation very much; however, present ideas on the uniformity of the universe are of recent vintage and are not accepted without difficulty. A new upheaval is always possible, but these ideas are now so firmly established in the theoretical conceptions of cosmology that if it is necessary to abandon them some day, then on that day, whether it be near at hand or still a long way off, the third age of cosmology will have come to an end.

We have already said that one of Hubble's discoveries, perhaps the most important of all, with which his name will always be associated, was totally contrary to his expectations, as it was to those of all his contemporaries, although it *was* consistent—it is very important to remember this—with the overall plan of regularity and uniformity within which he had conducted his research. I am referring to the spectral shift toward red of the light emanating from the galaxies, or the "red shift" of the galaxies (IIc3).*

This discovery was a total surprise to everyone, but it remains consistent with Hubble's cosmological scheme, in that red shift is a universal, regular phenomenon, obeying a relatively simple law and affecting all the images of galaxies in a uniform manner. On this point, even more than on the others, discoveries made subsequent to Hubble's death (1953) have confirmed the discoverer's views on the existence and characteristics of the phenomenon. The remote observer whom we have imagined would also see the spectra of the galaxies shift toward red, according to Hubble's law.

PATHS OF ACCESS TO THE REALM OF THE NEBULAE

Science as it is conceived and practiced in our time is not a system of findings that are fixed once and for all in an immutable form. And whatever the degree of certainty that can be attributed to its results, a description of these results is incomplete without some indication of the procedures and methods by which one arrived at them and which may enable one to correct them or go beyond them. This is particularly true in the case of cosmology, whose problems are on such a vast scale that it can scarcely lay claim to the same degree of certainty attached to a precise and limited result obtained in the laboratory. Moreover, before providing the details that are part of the picture of the contemporary cosmos, it would be suitable to give some indication of the sources of information and the methods of research, as well as the obstacles that had to be overcome to arrive at this picture. Everything that our science is capable of knowing beyond the moon it obtains by receiving and analyzing *radiation* which the stars emit, or, in certain cases, reflect and diffuse. For a very long time, astronomy has derived the totality of its information from

* À propos of this discovery, one frequently hears references to "the expansion of the universe." Red shift is a fact of observation, and the law which it obeys is an empirical law; but the expansion of the universe is a theoretical interpretation of this fact and this law. We shall examine the meaning and validity of this interpretation further on.

the visible radiation of light, and it is still this source that provides most of it. For the past twenty-five years, however, radioastronomy, which receives and analyzes the radiation of the stars on the wavelengths utilized by radio and radar, has contributed a valuable supplement to this information. More recently still, the astronomical study of X-rays has been initiated, for the stars also emit on these wavelengths.

First let us examine *light*, the optical domain which, in any case, determines the frame of reference. Three main techniques are generally combined in exploiting the images of galaxies:

- Studying the distribution of images of the galaxies on the celestial sphere, that is, on photographs that can be taken of it, particularly those taken with the aid of special large "panoramic" telescopes; this is a work of enumeration and statistical analysis.
- Photometric study of images of galaxies which enables us to judge their brilliance, their "magnitude," or the brilliance and magnitude of their various parts when they are sufficiently large; such studies are essential to the estimation, always difficult, of the distances of the objects being observed.
- Spectrographic study, that is, spectral analysis of the light coming from the galaxies.

The hypothesis of the homogeneity of the galactic system is confirmed by numerous and complex inferences based on investigations of the first and second types. The key to the problem lies in the distribution of the number of galaxies observed in terms of their luminosity, for on the average, the observed luminosity of a star is a clearly determined function of its distance.

But we must pause for a moment over the results of spectrographic research, from which cosmology has derived its most interesting results. According to the laws of classical and modern physics, the spectral analysis of light provides two kinds of information: (1) information about the chemical composition of the source, and possibly of the medium through which it has traveled (for atoms emit light over precise and characteristic wavelengths); and (2) information about the relative motion of the source in relation to the spectrograph, for this motion displaces the wavelengths.

When exploration of the realm of the nebulae began,

astronomy had already been taking advantage of these circumstances for a long time. The solar spectrum provides ample and extremely detailed information about the composition of the star; and many of the solar emissions that have been clearly identified are also found in the spectra of stars. The displacement of these wavelengths in relation to a reference spectrum having the same atomic elements enables one, by applying a well-known law, to measure the radial velocity of the stars in relation to the earth.* It was natural, therefore, to try to extend to the realm of the galaxies this spectrographic inquiry which was so productive in the world of stellar astronomy. We shall pass over the serious technical difficulties that had to be overcome in order to attain results. Among these is the red shift of the galaxies, which is certainly the most extraordinary and the most unexpected phenomenon that man, the experienced astronomer, has been permitted to see in the sky.

On the first point, the chemical composition of the stars, there was no very great surprise but, on the contrary, a confirmation of the classical hypothesis of the homogeneity of matter, a hypothesis which modern cosmology has transformed into a quasi-certainty. Indeed, the spectra of the galaxies reveal the wavelengths of the solar spectrum—less and less numerous and clear, of course, in proportion as the images become dimmer; but there is hardly room for doubt that the atomic composition of the stars and the stellar composition of the galaxies do not essentially vary, as far as present-day astronomy can tell.

It was on the second point, the investigation of the movement of galaxies by means of the spectrograph, that the great new revelation was made and that a seemingly inevitable and irreversible break with classical ideas about the universe took place. If one confines oneself to the study of galaxies close to our own, the results are not very different from what our knowledge of the stellar world might cause us to expect: certain spectra indicate a motion of withdrawal (shifting of the rays toward red), certain others, a motion of approach (shifting toward violet). This signifies, first of all, that the galaxy is turning upon itself; next, that

* This is what is referred to in classical physics as the Doppler effect (IIc3).

galaxies, like stars, are characterized by relative movements which are more or less disorderly and whose order of magnitude is not essentially different from that of stellar velocities. But as investigation is extended to dimmer images (and hence to more remote galaxies), shifts toward violet disappear. Only shifts toward red remain, and their amplitude increases in proportion as the images become dimmer as if after a certain distance galaxies were all impelled by a universal motion of flight, which becomes more rapid as they become more remote. Only "as if," for as we shall learn, this "motion" of expansion cannot be regarded as an ordinary motion, and its meaning can only be understood in terms of the leading ideas of the theory of relativity (IIIb1).

However, it is important to understand that even if one refrains from interpreting it as the effect of a motion, the red shift of the galaxies presents altogether exceptional characteristics. Indeed, as Hubble was able to prove in 1928*—and his results have since been confirmed for objects much more numerous and much more remote—the shift obeys a remarkably simple law which is valid for all images of galaxies, whatever their luminosity or their position on the celestial sphere. The shift in wavelength, or rather, the ratio between shift and wavelength, is a function of the "magnitude" of the image (that is, its luminosity); it is an increasing and simple function of this magnitude (IIc3). If Hubble's law is interpreted as describing a movement of expansion by the classical formula (which is valid, moreover, only at the first approximation), it means that the recession velocity of a galaxy is proportional to its distance, by which we understand that the movement of flight or recession is concealed in the case of the nearby galaxies by ordinary local movements. What is remarkable is the universality and simplicity of the law, which seems immediately to reveal to us a universal property of the galactic system and which confirms remarkably the impression of a system strongly united by universal structural properties, in short, of a *cosmos*.

* With the help of Humason, a former janitor at his observatory, who had become an eminent specialist in spectrography.

All this was acquired, then, by the traditional method of astronomy, the exploitation of optical data. And it is a cosmological frame of reference already determined in this way into which information derived from other sources has been and continues to be introduced. Among these sources, radioastronomy (Ib3; Id2–4) is, up to now, the only one that has contributed elements whose importance is undeniably cosmological.

From the cosmological point of view, in fact, radioastronomical research has to a large extent confirmed the image of a homogeneous distribution of matter in space. Sources of radio waves in the universe are numerous and varied. Some—stars or diffuse nebulae—are inside the galaxy, others are outside; it is the latter that are of interest to cosmology. More or less identifiable with galaxies or groups of galaxies, they are at any rate distributed on the sphere in a very uniform manner, and what can be conjectured about their distribution in depth makes it seem regular enough to be assigned, as we shall see, an important cosmological meaning. We also owe to radioastronomy the discovery of new objects which had eluded optical observation. The most spectacular was the discovery of quasi-stellar objects (Id2) (often called "quasars," for "quasi-stars") which optical telescopes confused with stars but which are not stars. We are still not sure what they are or where they are, but in all probability these objects are very remote and consequently of considerable interest to cosmology.

But the most recent discovery of radioastronomy, and probably the most important one for cosmology, is the discovery of a very weak radiation on the centimetric wavelengths (Id4) which, unlike radiations hitherto known, does not seem to proceed from any source we can identify. It is, in effect, isotropic, that is, identical in all directions, as if it were space itself that were radiating. It might be, as we shall see, a vestige of some state of the universe altogether different from the one we know.

In fact, however, in spite of rapid advances made in its techniques and aside from the exceptional case of this underlying cosmological radiation, radioastronomical research must in practice be associated with optical research in order for its cosmological implications to come to the fore; even in those cases where this association is possible, as in the case of quasars, it is not always easy.

THE COSMIC PICTURE

Let us try, then, to trace the main features of the cosmos as we see it almost fifty years after the discovery of the realm of the galaxies. Matter is composed of atoms which are everywhere the same, as classical scientists believed, and are divided into ninety-two naturally occurring species—325, if one takes account of slight differences in the nuclei. In some proportion not accurately known, but certainly substantial (at least half, perhaps, within the galaxies), atoms are combined into objects of more or less spherical form, namely the stars, whose dimensions and physical characteristics vary greatly, but which all have this common property of radiating, and notably of emitting light (Id4). The sun is an ordinary star in many respects; it is accompanied by a retinue of planets much smaller than itself which do not radiate. It is probable that other stars have planetary retinues, but up to now, no one has been able to observe one; however, this proves nothing.

The stars are not uniformly distributed in space—far from it. Many come in pairs, some in groups of three or four; there also exist larger, denser groups, which may vary from a few dozen to several thousand stars. On the average, the relative distances of the stars are very great in relation to their own dimensions; cosmic space is almost empty throughout. Stars move about in relation to one another at speeds which are high in relation to the speeds of human movement, but very low in relation to the speed of light (from a few dozen to a few hundred kilometers per second). Scattered stars, with or without planets, pairs, trios, and quartets, groups of varying size and density, matter not condensed into stars—all these elements are grouped together to form masses of much greater dimensions, within which the stars are numbered by the hundreds of millions, and even by the thousands of millions, namely, the galaxies. The sun and all the stars that we can observe with the unaided eye, with a few rare and precious exceptions, as we have said, belong to one of these systems, our galaxy (IIb).

Galaxies (IIc) vary greatly in shape and size. In general, they have a lenticular shape; often they exhibit a spiral form, with arms. In spite of these variations, they nevertheless constitute, like the stars, well-defined units that are structured in a very stable way by physical forces among which gravitation plays an essential role. Like the stars, they are very widely scattered; between galaxies, there is certainly more matter, stars or isolated atoms. The density of this matter (undoubtedly extremely slight) and its specific properties are still almost totally unknown.

In the scientific account (IIb–c), some precise information will be found on orders of magnitude in the world of the galaxies. From this it will be understood that our galaxy (whose dimensions far exceed anything that is imaginable to man—light, which takes a few seconds to reach us from the moon, takes more than a hundred thousand years to travel diametrically across the disc of the galaxy) is only an atom in the system of galaxies; one must multiply the diameter of our galaxy by ten thousand to arrive at the order of magnitude of that part of the system of galaxies that has already been investigated.

The fundamental problem of modern cosmology, from the point of view of observation, is therefore: is it reasonable to regard the universe as a fluid in which the galaxies are molecules? It is agreed so far that the answer may be yes, although this fluid is somewhat curdled. There is very little doubt, in fact, that galaxies, like stars, are rather regularly grouped into masses (IId). This is a very complex and rather controversial question, given the scale on which this grouping of galaxies occurs. At the present state of research it is generally accepted that our galaxy is itself part of a larger mass and that the groupings of images of galaxies on the sphere, which could be accidental, or even totally illusionary, actually correspond to real groupings produced by gravitation. Are masses of galaxies grouped in turn into masses of masses? Some people think so, but this is difficult to prove and these groupings of a higher order are not sufficiently established or systematic to validate the model of the cosmic "fluid" in any but an approximate sense.

The red shift and the law it obeys permit us to conclude—in a sense which we shall have to explain in detail in relation to modern theories which alone can provide a reasonable interpretation of it—that the cosmic fluid is in a state of uniform expansion, in which the distance between *any* two galaxies in the universe is an increasing function of time.

Galaxies are composed of stars which radiate as long as they exist, that is, for periods of time which may extend to thousands of millions of years. This radiation, which is unstable and temporary by its very nature, is nevertheless, along with matter—at the expense of which, in fact, it is produced by the stars—a permanent component of the content of the universe. Besides, as we have said, it is a widely verified consequence of the theory of special relativity that matter and radiation are two forms of energy which are quantitatively equivalent and, up to a certain point, transformable one into the other. The quantitative comparison of the content of the universe in matter and its content in radiant energy is very important for cosmology.

One of the remarkable features of the universe that we observe is, first of all, that light (and, in general, radiation) is rare in relation to matter, a circumstance which may be regarded as contingent in that there is nothing in known physical laws, either classical or modern, that makes it necessary or even probable that there should be a certain proportion between matter and radiation. A "universe of light," in which there would be nothing but radiation, is just as conceivable, in theory, as a "universe of matter," in which there would be nothing but matter, a case much closer to the universe that we see. But under certain physical conditions the proportion could be quite different, and it could have changed appreciably during a former period of cosmic history.

THE PROOF OF A COSMIC HISTORY

The phenomenon of the red shift and Hubble's law also reveals in a particularly striking way another essential feature of the universe which neither the ancient nor the classical cosmology had introduced into its descriptions: namely, that the universe as a whole is undergoing evolution. Herein lies a kind of paradox which deserves brief reflection, for the idea that "everything changes," that "everything passes," that "the past never returns," etc., is one of the most ancient truisms of Western thought which philosophers and poets have repeated and commented on copiously for twenty centuries and more. And yet, until our own time, cosmology—to say nothing of metaphysics* —had always eliminated this truism from its systematic descriptions. The ancient world was cyclical; the classical world, on the other hand, was static: endowed with an unchanging geochronometric form, it was assumed more or less implicitly, and sometimes explicitly, to be composed of indestructible elementary objects impelled by unceasing but random movements whose nature was such that *on the average,* within large spaces and over long periods of time, everything remained in the same state. It is true that after the middle of the nineteenth century the expounding of the second principle of thermodynamics (IIIb) had led some physicists to raise, albeit in somewhat indefinite terms, the question of a cosmic history; but in general, their intention was more often to deny it, in spite of the second principle.

The red shift of the galaxies provides something that had always been lacking, something that gives the truism of "becoming" a cosmic dimension: namely, a phenomenon of total evolution, involving all of the observed universe. Indeed, however one interprets the red shift—as a kinematic effect that is properly speaking relativistic (which is the most reasonable and most usual interpretation), as a classical "movement" of expansion, or even (although this interpretation is very unlikely) as a modification of light occurring in a static world—red shift always signifies that the universe as a whole is undergoing a transformation which is governed by a well-defined law.

Classical astrophysicists were not unaware of the fact that, contrary to what the ancients had thought, the truism of "becoming" must also be valid for the stars. But in the first place, they had no means of interpreting the history of the stars, observing simply that the enormous output of energy of the sun, prolonged over a long enough period of time to include at least the whole

* The shameless "eternalism" of classical philosophy may explain how philosophers like Bergson acquired part of their fame by placing at the center of their thought a truism as worn out as the idea that "everything passes," without having the slightest element of serious reflection with which to approach the problems it conceals.

history of life on earth, was a mystery that was completely insoluble for classical physics. Second, in the case of the movements of the stars regarded as a whole, classical astrophysics applied with mathematical care and exactness the "classical" hypothesis which we have reviewed: that these movements are sufficiently disorderly* so that, on balance, they compensate for one another and that their average may be used to define a system of coordinates which is at rest in relation to absolute space. This hypothesis was so familiar to classical thought that we shall find it again, duly modified to accord with relativistic principles, in the first cosmological study of Einstein, although this work inaugurated the modern era of theoretical cosmology.

In the first half of the twentieth century, and particularly in the second quarter, thanks especially to the rise of quantum physics and nuclear physics, the processes which determine the radiation of the stars—a fundamental phenomenon in the universe—and which therefore condition their evolution, were discovered. Thus it was established that the history of the stars, and hence the history of the universe, is not separable from the history of atoms, for it is the transmutation of hydrogen, element number one in Mendeleev's classification, into helium, element number two, that furnishes the most substantial part of the radiant energy of the sun and the stars (IId2, 4). Painstaking research, which is still in progress, next showed that other nuclear transmutations involving heavier nuclei also occur in stars. With these discoveries, history was penetrating into the very texture of things, for the fusion of hydrogen into helium, no matter how it is done (there are several ways), is a striking example of an irreversible process, precisely because a portion of the initial mass is transformed into radiation. In order to "regenerate" hydrogen, it would be necessary to have physical conditions which do not *currently* exist in the universe.

Here we may be touching on one of the most profound points of rupture between the classical and the modern philosophies of nature. In the mechanical models of classicism, the atom or the fluid, endowed with inertia, was unchanging. Events re-

* This hypothesis does not contradict the one according to which the movements of the stars obey very rigorous laws; the two are superimposed in accordance with order of magnitude, without this creating any ambiguity.

sulted from its movements, but left its other attributes unchanged, so that if an event occurred, one could always conceive of an opposite event restoring things to their previous state. The modern conception of matter rules out the idea of a physical substance that is indefinitely identical to itself. It makes the opposition between the event and the thing a matter of relativity, and this accords well with the new conception of space and time which forms the basis of special relativity and which substitutes for space and time a space-time *continuum* made up of point-events (i.e., "things" and "events" do not exist in themselves, but are defined relative to one another).

The atoms and the stars, and along with them the planets and the galaxies (although our knowledge of the evolution of the latter is still rather vague), seem, therefore, to be engaged in processes of partial evolution which, although of course distinct and relatively independent from one another, are all irreversible and are integrated into a single evolution, a common history. This impression is remarkably well confirmed by the estimates that can be made of the approximate duration of some of these partial processes. For example, an analysis of the radioactive minerals on the earth enables one to arrive at an approximate date for the formation of the terrestrial crust. Astrophysicists, working with entirely different methods, are able to give an estimate of the life span of the sun and of stars belonging to other types, and also to judge the age of certain masses of stars in the galaxy, and even of the galaxy itself.

Now, it is remarkable that these measurements (*a*) agree relatively well among themselves, since one is always dealing with a minimum of a few billion years and a maximum of a few tens of billions of years, and (*b*) all result in *short* durations.

Naturally, in relation to the life of a man, or even to the whole history of civilization, ten billion years is a period of time that does not seem "short." (It is five million times the interval that has elapsed since Christ.) But if one considers that the earth's crust may be five billion years old, that *life* on earth may be several billion years old, and that the light that reaches us from the most remote galaxies was emitted several billion years ago, this order of magnitude will seem less improbably long.

However, the important point is that this order of magnitude agrees with the one that can be conjectured for the evolution of the universe as a whole by using the red shift of the galaxies as a standard of measurement. Without attempting to go into the theoretical foundations of this calculation (IIc3), we can show its principle in a relatively simple and intuitive way. If red shift is the result of a relative movement of the galaxies, an expansion of the universe, observation and Hubble's law enable us to calculate a coefficient of proportionality between the distance of a galaxy and its speed of recession, Hubble's constant, or H.

Let us assume that this expansion is uniform, i.e., that the distance between any two galaxies increases proportionally to time. (This is a working hypothesis which, naturally, is not at all self-evident, although it is almost exactly correct over a long period of time.) Therefore, at a certain moment in the past, these two galaxies must have been in contact (and somewhat earlier, it was their stars that were in contact). Therefore, simple reasoning and calculation show that (*a*) *all galaxies* must have been in contact at that moment; and (*b*) one need only invert H to find the date of this encounter, $T = 1/H$.

Although measurement of H remains difficult and rather imprecise, it is nonetheless possible. The most recent estimate, based on data obtained on Mount Palomar, is $T = 13$ billion years (approximately). In fact, this length of time is of the same order of magnitude and, generally speaking, greater than the estimated durations of the great astrophysical processes.

T is sometimes referred to as the age of the universe, but this expression is inappropriate for two reasons. First, it is by no means self-evident that one can regard the moment when all the galaxies and the matter they contain were in contact as the "birth" of the universe. Second, we do not know with certainty whether the expansion is uniform, and thus whether this great concentration occurred at this date or before or afterward, or even whether it occurred at all. It nevertheless remains true that the constant H, on the basis of observation, at the very least provides a cosmic scale of time, and it is quite remarkable that the durations of all the great astrophysical processes that we know agree with it reasonably well.

Moreover, for all kinds of reasons which it would take too

long to enumerate, the idea of a universe evolving from an "initial" supercondensed state, however extraordinary it may be, and despite the many and lively criticisms it has provoked, has had considerable impact, and there are, at the present time, few astrophysicists who would deny it totally and without reservation. The most astonishing thing is that at the moment we have at our disposal no means of learning what this supercondensed state may have proceeded from; nor do we know the cause of this species of explosion, of which the present form of the universe might be the remote consequence and of which the background radiation I mentioned might be a vestige. Was it preceded by a contraction that could be regarded as "symmetrical" with the expansion we are witnessing? Unfortunately, if we run the film backward, so to speak, what we know about the laws of physics and the structure of matter does not enable us to explain how the contraction stopped and the universe surged outward. We are dealing here with what mathematicians call a "singularity," that is, an exceptional state of things in which the formulas cease to have any meaning. Do not ask a geometer to tell you the tangent of an angle of 90°, because for this value of the angle, the tangent is not defined. Similarly, if you ask a cosmologist to tell you the density and temperature of the "cosmic fluid" at its origin, he will tell you that by definition the origin is a point in time when these quantities were not defined. The curious thing is that everything conspires to make one think that the universe found itself in conditions such that it could not avoid passing through this "singularity."

IV. THE THEORY OF THE UNIVERSE

In our review of the discoveries of cosmological observation, we have long since left the paths of theory. It is high time we returned to them, for without direct and precise reference to the notions of the theory of relativity, anything that can be said about the modern universe remains approximate, just as the elementary, atomic and subatomic mechanisms that determine the evolution of the stars remain incomprehensible within the classical frame of reference.

PRINCIPLES AND QUESTIONS OF THEORETICAL COSMOLOGY

Let us review a few essential ideas—some principles, others consequences—of the theory of relativity:

▪ Measurements of space and time are separable only in relation to a given observer; objectively, they are inseparable. What is invariable, independent of the observer, and hence absolute, is therefore neither the distance between two points in space nor the interval of time between two events, but the world interval between two point-events.

▪ Matter—inert and characterized by mass—and energy, especially radiation, are merely aspects of a single reality all of whose forms can be transmuted into one another in such a way that in each of its transformations a single dimension, matter-energy, is quantitatively conserved. These two statements are common to the theory of special relativity and the theory of general relativity.

▪ Forces of inertia and forces of gravitation are of the same nature, and it follows from this that the metrical properties of space-time depend on the distribution of matter-energy in the vicinity of the spatio-temporal region in which these metrical properties prevail; this statement is peculiar to the theory of general relativity.

From the point of view of cosmological theory, it is the third statement that is most important. In effect, the formal cosmology of classical physics, which was more or less superseded by the theory of special relativity, assumed *a priori* a geochronometric form that was universal, homogeneous, and regular. But general relativity, for which spatio-temporal measurement is primarily "local," cannot envisage the existence of such a universal and all-embracing form except in terms of the distribution of matter-energy on a cosmic scale. The question arises as to whether there exists a distribution having properties that are significant on this scale, without which there would be a multiplicity of geometries whose agreement would not necessarily be defined.

Discoveries based on observations outside the galaxy are decisive in this regard. Since matter—and with it, radiation—is uniformly distributed, there must be a uniform and regular geometry that corresponds to it, whatever local distortions of this geometry may be caused by irregularities in the local distribution of masses. Moreover, the fact that with the exception of electromagnetic or nuclear forces only the mechanical forces of inertia and gravitation actually enter into the relativistic equations, has no disastrous consequences for cosmology, for the interactions of gravitation are dominant on a very large scale.

By regarding matter-energy as a "cosmic fluid," it is possible to find universal, cosmological solutions to Einstein's equations which determine the relationship between matter-energy and geometry (IIIa2, 3). What, then, is this universal geometry, or rather, this universal geochronometry? Let us begin by noting that, in the present state of knowledge, the problem does not have *one* clearly defined solution. Several solutions or, as is often said, several "models of the universe" are conceivable, but what can be reasonably asserted is that *the* solution, whether or not it will one day be accessible, belongs to a certain family of models possessing well-defined properties.

In order to understand the characteristics of this family (IIa3) more clearly, let us disregard for the moment the problem of time and focus our attention on the properties of space. We shall see whether, why, and up to what point such an abstraction is acceptable. If the "cosmic fluid" really has those properties that observation suggests it has, then the geometry obtained resembles traditional Euclidean geometry in that it is uniform and regular. The "curvature" of space, instead of being zero, is constant. (In fact the geometry is actually Euclidean in certain models.) If the constant "curvature" is positive, then space is "spherical"; it is closed and finite, although it has no boundary, as is the surface of a sphere when reduced to its two dimensions. If the constant "curvature" is negative, space is "hyperbolic" and infinite like Euclidean space, but the equivalent that one can give of it in two dimensions is less simple to describe than the sphere and not as perfect.

But now what has happened to the relation between space and time? Have we the right to separate geometry from chronometry as we have just done? The question is not simple and the answer is both yes and no. It is no because Einstein's equations define the metrical structure of a space-time in which the metrical properties of space are not *a priori* independent of time—and indeed they cannot be in a physical universe in which we observe a red shift. It is yes, however, because in this space-time corresponding to the observed cosmic distribution of matter-energy (IIIa1), a certain family of "time lines" can be distinguished automatically, and because on each of these lines one can measure a "cosmic time" that is common to them all. This results, in our universe, from the properties of the galactic system and from the characteristics of red shift. In fact, the latter depends only on the distance of the galaxy in question and not on the direction from which it is being observed. This proves that galaxies are characterized by a universal "motion" of flight which attains enormous speeds in the case of very remote objects, and that their relative local movements have low speeds. This allows us to make an analogy between the spatio-temporal paths of the centers of galaxies and the family of lines of time that measure "cosmic time." Space thus becomes a part of space-time which is universally separable at any moment, and which can be regarded as a clearly distinguished geometric form. (Its "curvature" is distinguished from that of space-time; in certain models it is zero. Space is Euclidean, but the curvature of space-time is not: space is undergoing expansion.)

This brings us to the relativistic interpretation of the red shift of galaxies (IIIb1). This effect is seen, not exactly as the consequence of a "motion," but as resulting from the special way in which space and time are combined in cosmic space-time. It expresses a kind of dilation of space, for any two time lines of the fundamental family of lines are moving away from one another. It follows from this that the wavelength of the light that joins two point-events situated on two lines of this family is, upon reception, displaced in accordance with a law *analogous* (and only analogous) to the classic Doppler effect.

All this was achieved, not without difficulty, between 1917 and 1927, thanks to the efforts, rather scattered in the beginning, of Einstein and a few mathematicians, physicists, and astronomers who were capable of following him down these new and revolutionary paths—the Dutchman Willem De Sitter, the German Weyl, the Russian Friedman, the Englishman Eddington, and the American Robertson. In 1927, while Hubble and the spectroscopist Humason were completing their work establishing the red-shift law at Mount Wilson, the Belgian priest Georges Edouard Lemaître was the first to demonstrate explicitly that the shift, a phenomenon virtually unintelligible in terms of classical physics, was perfectly natural and predictable in terms of a relativistic cosmology.

This was a great triumph for theoretical thought, which nothing to date has seriously threatened. The classical pseudo-cosmology, a dubious mixture of *a priori* ideas and incomplete observations, was now replaced by a real cosmology in which, thanks to a combination of observation and mathematical theory, the universe—regarded as a whole, form and content closely associated—became once again an object that could be known and described. Thus modern cosmology revived the ancient sphere, on the grand scale of modern telescopes and with the aid of the most refined mathematics.

But it is inconsistent with the spirit of modern science to become fixed in the contemplation of the results it has achieved—no matter how satisfying to the mind these may be—and the inventors of the new cosmology, like all their successors, have primarily concentrated their attention on the new problems raised by this new image of the universe. Broadly speaking, one can divide these problems into two groups which correspond to two principal questions: (*a*) Is it possible to determine *completely* the geochronometric structure of the universe by means of theory and observation, and if so, how? (*b*) Must we accept the idea of a singular origin of the universe (or at least of the present period of its existence)? Can one avoid this idea, and if not, how are we to understand it?

The two questions are not independent. The first, more theoretical and in some sense more esoteric, has often been obscured by the second and the polemics to which it has given rise have lost none of their sharpness after forty years; but this question should not be ignored.

▪ Ever since the first victories of modern theoretical cosmology—and nothing has happened since then to seriously threaten this hypothesis—it has appeared that, in view of the results of observation, the structure of cosmic space-time—or the model of the universe representative of our world—should, as we have said, belong to a clearly defined family. But in order to go further toward a true definition of the model, two unknown elements must be determined. These are a certain numerical factor, which we shall call k, which may be equal to -1, 0, or $+1$, on which the structure of *space* depends,* and a certain numerical function of cosmic time t, which we shall call R, or rather the behavior of this function over a rather long interval, on which the temporal behavior of the universe depends. Provided we give Einstein's equations a form that is not absolutely general,† in accordance with Einstein's advice (he gave lengthy consideration to this question), the two unknown parameters k and $R(t)$ may, in principle, be determined by observation. But the degree of precision that has up to now been achieved in the measurement of remote galaxies is not sufficient to produce any definite answers. In particular, in spite of the recent completion of some new and very ingenious tests, it is still impossible to decide whether cosmological space is hyperbolic ($k = -1$), Euclidean ($k = 0$), or spherical ($k = +1$). Red shift indicates at once that function $R(t)$ is increasing, that is, that the universe is in expansion, but there is still insufficient information about the second derivative of R, on which the deceleration or acceleration of its expansion depends. This expansion is certainly almost uniform, and may be slowing down very slightly. It is important to determine this; if the theory should be confirmed, this would signify that expansion was more rapid in the past, and this would only give more importance to the question of the original singularity, which would then move closer to us in time. It is not out of the question, furthermore, that the function R is periodical or quasi-periodical and that as a consequence the universe has undergone

* We have noted above that, in the space-time models of the cosmological family, space, on the one hand, and a cosmic time, on the other, may be objectively distinguished in space-time.

† That is, by canceling out the "cosmological constant" in these equations (IIIa2, equations 1 and 2; IIIa3, equations 3 and 4).

and must repeatedly undergo this singular state of which certain vestiges may still be visible. We see that the myth of the "eternal return," often associated with the truism of "becoming," also has an unexpected counterpart in contemporary cosmology.

While awaiting the time when the moon or some orbital station can accommodate a large telescope which, freed from the atmospheric screen, would penetrate much more deeply into remote space (or rather into remote space-time, for observation travels backward through time), many cosmologists accept as a frame of reference and a basis for calculation a model once proposed by Einstein and De Sitter; it has the merit of being very simple (IIIa4 and d). In this model, $k = 0$—space is Euclidean—and the function R is expressed $R = At^{2/3}$, A being a constant. The interval that separates the present cosmic age from the initial singularity is, in this model, exactly equal to ⅔ of the inverse of Hubble's constant, or about nine billion years. This gives us a universe which is a little too "young," but on the whole the model agrees rather well with the results of observation.

THE CONTROVERSY OVER "ORIGIN"; IMPROBABILITY OF THE STATIONARY UNIVERSE

The question of the *initial singularity*, the "big bang," as English-speaking writers sometimes call it so expressively, ultimately dominates all the others, whether one approaches the matter from the standpoint of observation—as we have done—or examines its theoretical implications. When one calls to mind the whole development of cosmology for the past fifty years, it is obvious that this is the outstanding question. Ever since cosmologists perceived it as such (around 1930), there has always been a kind of division among them between those who accepted it and used it as a basis for ideas and research and those who looked for ways to avoid it. At one extreme, for example, we find Lemaître and George Gamow; at the other, Hermann Bondi, Fred Hoyle, and supporters of "the steady-state theory." Let me say at once that recent theoretical research has shown that within the frame of reference of relativistic cosmology, one cannot avoid solutions characterized by singularities. In the equations of Einstein and of reasonable and very far-reaching hypotheses on the cosmic matter-energy, all solutions involve one or more singularities. If one wishes to avoid them, one must go outside general relativity and construct a cosmology which is relativistic in a sense, but which is not based on the equations of Einstein. This is an audacious enterprise which is, moreover, from the outset, independent of the problem of singularity. (The first person who attempted it, the Englishman Milne, *believed* in the singular origin of the universe, to which he even gave the theological sense of a *creation.*) Such an enterprise may be understood in many ways; we shall confine ourselves to mentioning the most notorious of the attempts that have been made to construct a cosmology of this type, because its consequence was simply to eliminate all question of origin. Its inventors were two English mathematicians, Bondi and Thomas Gold; its principal defender was an astrophysicist, also English, Hoyle.

This steady-state theory deserves particular attention, in any case, first because of the originality and force of its logical procedures, and also because of the controversies—it would be better to say polemics—to which it has given rise. Its authors, breaking totally with an empiricist conception more or less common to physical science (the discovery of laws results from an accumulation of experiments), are of the opinion that cosmology would proceed *a priori.* In effect, they say, the condition for the existence of any physics is that its experiments be repeatable everywhere and always, that its laws may, in principle, be verified by any observer, at any time; this presupposes the existence in the universe of a metrical framework that is uniform and regular. Indeed, this is what happens in relativist cosmology, when it chooses a certain family of models possessing, precisely, certain very characteristic properties of regularity. But whereas relativistic cosmology looks to observation—in vain, as it happens—for the means to completely determine the model, the steady-state theory, on the contrary, proceeds *a priori,* by drawing all the consequences from the *perfect cosmological principle* whereby *for any observer* and *at any time,* the universe, considered on a large scale, must have the same overall appearance.

It is surprising to observe that from such an abstract principle, its authors draw some very precise conclusions about the structure of space-time, to the point where they completely determine their model of the universe. Let us examine its properties: it is from the first incompatible with the theory of general relativity. It is not, in fact, a solution to Einstein's equations, except for the unreal case of a totally empty universe.* The universe is in expansion, and is not therefore static, but it is steady in the sense that in spite of its expansion, the appearance of the universe is renewed in such a way that apart from constant alterations in details, any observer sees and would always see the same thing on a universal scale even if he pursued his observations indefinitely.

This leads us to one of the most original and most debated aspects of the universe of Bondi and Hoyle. The expansion causes all galaxies to recede; for any observer, this phenomenon of recession would make them gradually disappear "at the horizon."† The universe ought therefore gradually to become empty before his gaze—but this consequence is contrary to the perfect cosmological principle. It is therefore necessary that new galaxies constantly appear to replace those that disappear, that is, that everywhere and always, new matter arise *out of nothing* in space. This is the hypothesis of continual creation.

It is a hypothesis that is manifestly contrary to the principle of the conservation of matter-energy which modern physics has inherited from classical science and to which it has given a more general application. It is true that quantum physics talks about the "creation" and "annihilation" of particles, but this means that a particle is formed out of a radiant energy already present, or is transformed into such an energy. In the steady-state theory "creation" occurs out of nothing, as if existing matter were continually conducting new particles into its vicinity. "Magic," say many physicists, who refuse to renounce the most fundamental principle of their science; but supporters of the steady-state theory reply, in substance, as follows. It makes no sense to say that matter-energy is conserved in its totality and absolutely in

* This is the "De Sitter's universe" (IIIa4 and b).

† They do not *leave* the field of observation, but their luminosity tends toward zero.

an infinite universe like that of the steady-state theory, for in such a universe this quantity is not defined. For the principle to have any meaning, therefore, it is necessary to state specifically in relation to *what* matter-energy is conserved. Our hypothesis is only one of the possible ways of stating this, and it is a way that is consistent with the perfect cosmological principle: matter-energy is conserved within the field of any observer. It should be added that, when one considers the dimensions of the universe being observed and the rate of expansion, this continual creation would take place so slowly that no laboratory experiment could cause it to appear.

However this may be, the steady-state theory has the great merits of lending itself well to controlled observation, for its principles enable one to deduce numerous phenomena that are in principle observable, and of being subject to the test of new phenomena. Leaving aside the consequences it has in common with other cosmological theories, we note, for example, that it allows one to foresee an acceleration of the expansion at a precisely determined rate. We also note that any proof of a similarity between near and distant objects is favorable to it, and on the other hand any indication of a systematic difference between these is unfavorable to it. Anything that is observed must, according to the steady-state theory, be the result of the "present," "daily" activity of the stars and galaxies; inversely, any phenomenon that presents itself as a vestige is, according to the theory, incomprehensible.

The steady-state theory is twenty-five years old; for the last few years, advances in observation have been turning distinctly to its disadvantage, to the point where some of its supporters have openly abandoned it. The fact that the factor of deceleration of the expansion has been estimated to be in the vicinity of 0.2, whereas the steady-state theory sets it at -1 is not in itself very significant, given the uncertainties that make measurement imperfect. But a certain number of findings, none of which is decisive, have been accumulated against the steady-state theory, creating a growing impression of improbability. The distribution of radio sources in terms of their "loudness" indicates that their number clearly increases more with distance than the steady-state theory allows one to foresee. This is an indication that these are astrophysical formations (essentially, galaxies) which are emitting radio waves at a certain period of their history, a period which has come to an end in the case of objects which we observe in our vicinity, that is, objects at a more advanced period of their intrinsic evolution. The background radiation to which we have referred can be naturally interpreted as the vestige of a state of the universe in which the radiation was much more intense; this phenomenon is very difficult to interpret in terms of the steady-state theory.

Yet another discovery has come along recently to diminish the credibility of the steady-state theory: the abundance of helium in the galaxy is too great to result only from what the stars produce. One must therefore assume the existence of a certain quantity of helium which is to be found in the stars, not because it was formed there from hydrogen, but because it was already present in the diffuse matter out of which the stars were formed. Where did it come from? Here again, it is extremely tempting to assume that it is a vestige of the supercondensed state, in which nuclear transmutations occurred under exceptional conditions. Other arguments, particularly those drawn from the properties of quasars, also tend in the same direction, although they are less compelling.

The present situation is hence more favorable to the other tendency, that of the cosmologists who accept the idea of the original singularity and who define their hypotheses in relation to this idea. The first person who pursued this path, a long time ago, was Lemaître. In 1931 this clergyman and mathematician who had broken with the methods of the theory of relativity and was fascinated by the quantum idea then taking hold in microphysics, conceived of the origin of the universe as a genesis in the course of which there appeared out of a unique quantum not only matter and energy, broken down into atoms and quanta, but also the form of this atomization itself, space-time.

There was more dialectics than physics in the "primitive atom" of Lemaître; but fifteen years later Gamow and some other eminent specialists of nuclear physics (in the interim, the events at Hiroshima and Nagasaki had demonstrated the true seriousness of this new physics) took up Lemaître's idea, basing their work on much more solid physical foundations. Imagining the cosmic matter a few seconds after the original singularity as composed solely of the simplest microphysical particles in a state of extreme condensation, at a fantastically high temperature but in process of a very rapid expansion and cooling, they attempted to reconstruct the gradual synthesis of the atomic species, then the stabilization of the mixture obtained in proportions close to those now observed. It was a considerable piece of work, the first attempt in the history of science to describe the genesis of atoms; it was also the first serious effort by atomic physics to enter the realm of cosmology.

In the years that followed the steady-state theory dominated the scene, but the great work of Gamow had shown that the synthesis of atomic species out of hydrogen could only occur, beyond helium, under altogether exceptional physical conditions, and indeed, the very ones that may have prevailed around the time of the singularity. Could the steady-state theory, which rules out singularity, explain the synthesis of the atomic species out of the new matter whose continual creation it postulated? It would really have been too easy to assume that this matter is formed by respecting precisely those proportions which we observe between the atomic species. Hoyle and three other physicists undertook to demonstrate that synthesis of all the atomic species may take place in the centers of stars; they succeeded in demonstrating that the genesis of complex atoms out of hydrogen was at least conceivable in the centers of stars in a stationary universe, which, from the point of view of the steady-state theory, was a result of the first importance. But since then, as we have said, the steady-state theory has, for other reasons, lost much of its credibility, and research on singularity has resumed with greater force, enriched by the advances achieved since Gamow's work in the understanding of nuclear phenomena. It should be pointed out that Soviet astrophysicists, long absent from the cosmological scene for complex ideological reasons, are now participating actively in this type of research.

Quite recently a French physicist, R. Omnès, has undertaken research on the period immediately surrounding the initial state based on the hypothesis that *a priori,* particles and antiparticles —matter and antimatter—could be equally abundant. This hypothesis had in fact already been considered by some Swedish astrophysicists,* but in a very different cosmological context.

COSMOLOGY AT THE FRONTIERS OF SCIENCE

What is particularly striking about the studies we have just mentioned and what makes them so characteristic of the renascent cosmological spirit of the twentieth century is not only—as, after all, we might have expected—that they employ all the theoretical and experimental disciplines related to exploration of the physical world, but also that they ignore in a surprising and indeed almost incredible way the distinction between orders of magnitude which is so familiar to modern physical methodology. The theory of special relativity in which the overall history of the universe is interpreted describes phenomena on a very large scale. It is well suited, in extension and duration, to the orders of magnitude of astronomy. (The first *laboratory* test of this theory was carried out, successfully, in fact, in 1960, or forty-five years after its creation, whereas it had received its first confirmation from astronomy in 1919.)

Nuclear physics, on the contrary, describes interactions that take place in areas on the order of one hundred billionths of a millimeter and on time scales in comparison with which the few minutes of the "life" of a neutron are a veritable eternity. As a result one must count in seconds the duration of those "initial" processes which may have determined the history of the entire universe for billions of years. In physical science, orders of magnitude are superimposed: something that is infinitesimal from one point of view becomes enormous from another; but cosmology places durations of totally different orders of magnitude end to end. Something that happens in a tenth of a second is, generally speaking, altogether negligible outside certain very limited areas (the life of a man or that of a huge summer cloud). But around the time of the cosmic singularity, a tenth of a second was an im-

* Notably Hannes Alfven who, together with Néel, won the Nobel Prize for Physics in 1970.

portant period of time for the whole course of the history of the universe.

This contempt for the distinction between orders of magnitude is a good illustration of those differences which, at one point or another, inevitably appear between the spirit of cosmology and the spirit of physics, in spite of the essential role played by the laws of physics in the construction of cosmology. The very question of singularity, however briefly one analyzes its implications, clearly manifests this difference. Let us mention only one of these implications. Suppose someone asks what there was before the singularity. Everything that has happened *after* it is conceived according to the hypothesis of an irreversible evolution; to *arrive* at the singularity, it would have been necessary for all the processes to occur *in reverse*, but can we conceive that, in two *successive* segments of a *single time*, two total evolutions that were exactly the opposite of one another could have occurred? (When one runs a film backward, this takes place in a world which itself continues to turn *right side out*, but in the cosmological case, the film is not distinguished from the rest of the world.)

In modern physics, as we have said, time is not, as in Newton's system, an empty and prefabricated structure within which events take a place prepared in advance; an instant must be marked by an event, two successive instants by two successive events, and the direction of time must be marked either by a fundamental sequence or by a property common to all sequences. But if *everything* is reversed at the instant of the singularity, how can we mark by two *successive* events an instant *before* and an instant *after* the singularity? Must we rather say that the contraction has not *preceded* the expansion, but that it is in reality an image in which the past and the future are reversed as the right and left sides are reversed in a mirror?

If the reader regards this question (and many others of the same kind, which cosmology inevitably leads one to raise) as totally futile and unworthy of true scientific consideration, or if he finds that it puts him in a state of philosophical insecurity incompatible with the serenity of rational work, then we have not succeeded in convincing him that cosmology is really a science.

MODERN SCIENCE AND THE UNIVERSE

Scientific Account

I. THE MAJOR IDEAS OF MODERN PHYSICS

a. THE THEORY OF SPECIAL RELATIVITY

a1. REFERENCES OF INERTIA IN CLASSICAL MECHANICS. In the second part we defined a Galilean reference or an inertial reference. We saw that two such references have, in relation to one another, a motion of rectilinear and uniform translation. Let us refer one of the references to three axes of rectangular coordinates Ox, Oy, Oz, and the other to three axes O$'x'$, O$'y'$, and O$'z'$. One can always choose these axes in such a way that O$'x'$ coincides with Ox, O$'y'$ being parallel to Oy and O$'z'$ being parallel to Oz. We shall assume this to be the case in all that follows. Let us call the unit consisting of a point in space M and a time t the "event"; the coordinates of the event will be the three coordinates of M along

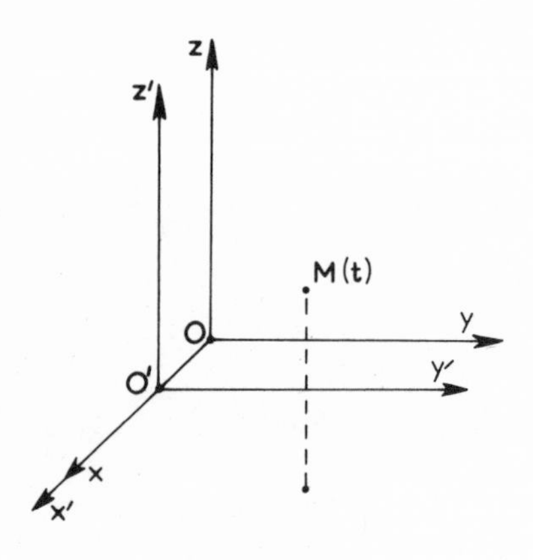

FIGURE 55

one of the systems of axes and time t (Fig. 55). Let x, y, z, and t be these coordinates at reference $Oxyz$ (reference E). At reference $O'x'y'z'$ (reference E′), the coordinates of this event will of course be

$$x' = x - vt, \quad y' = y, \quad z' = z, \quad t' = t \tag{1}$$

v being the constant speed of point O' along Ox. Now let us consider another event whose coordinates are $x + dx$, $y + dy$, $z + dz$, $t + dt$. This gives us, based on the preceding formulas,

$$dx' = dx - vdt, \quad dy' = dy, \quad dz' = dz, \quad dt' = dt$$

dx', dy', etc. being the increments that must be added to coordinates x', y', etc. to obtain the coordinates of the new event at reference E'. If this second event is simultaneous with the first, dt is zero, hence $dx' = dx$, $dy' = dy$, $dz' = dz$, and consequently

$$\overline{ds}^2 = \overline{dx}^2 + \overline{dy}^2 + \overline{dz}^2 = \overline{ds}'^2 = \overline{dx}'^2 + \overline{dy}'^2 + \overline{dz}'^2$$

This expresses the fact that the distance between two points is constant when one passes from one system of coordinates $Oxyz$ to another $O'x'y'z'$. Similarly, $dt' = dt$. We see that one can separate the spatial coordinates x, y, and z from the temporal coordinate time t, since the latter coordinate is independent of the spatial coordinates.

Let us suppose that, during the very short interval of time dt, point M passes from the point of coordinates x, y, z to the point of coordinates $x + dx$, $y + dy$, and $z + dz$. Its speed $\vec{V}$ in relation to reference $Oxyz$ is, as we have seen, vector $\vec{V}$ of components $\frac{dx}{dt}$, $\frac{dy}{dt}$, and $\frac{dz}{dt}$; its speed in relation to reference $O'x'y'z'$ will be, according to the formulas for changing coordinates,

$$\vec{V'} = \vec{V} - \vec{v}$$

$\vec{v}$ being the vector of components v, O, and 0 on axes $Oxyz$, that is, the vector velocity of O' in relation to $Oxyz$.

One can write the preceding relationship $\vec{V} = \vec{V'} + \vec{v}$. This formula expresses the very simple law of the composition of speeds; this law is perfectly compatible with the existence of speeds as high as is desired. In particular, if $\vec{v}$ is not zero, that is, if the two spatial references $Oxyz$ and $O'x'y'z'$ are not at rest in relation to one another, $\vec{V}$ and $\vec{V'}$ are always different, except when one of these two speeds is infinite, for in that case the other is also. Thus infinite speed plays here a special role of absolute speed which does not change when one passes from one inertial reference to another.

However, in reality, the speed of light in a vacuum, which is not infinite but is equal to approximately 300,000 kilometers per second, possesses precisely this property of always being exactly the same when one passes from one Galilean reference to another. This has been demonstrated experimentally, in particular by the famous experiment of Michelson and Morley.

A pencil of light issuing from O (Fig. 56) is divided in two by a semitransparent mirror M. Ray OAM_2 is reflected at A, reaches M_1, then, reflected at M_1, reaches I. Because of the different paths traveled by the rays, bands of interference are produced at I. One of the rays was parallel to the vector velocity of the earth in relation to the sun, the other was perpendicular to it,

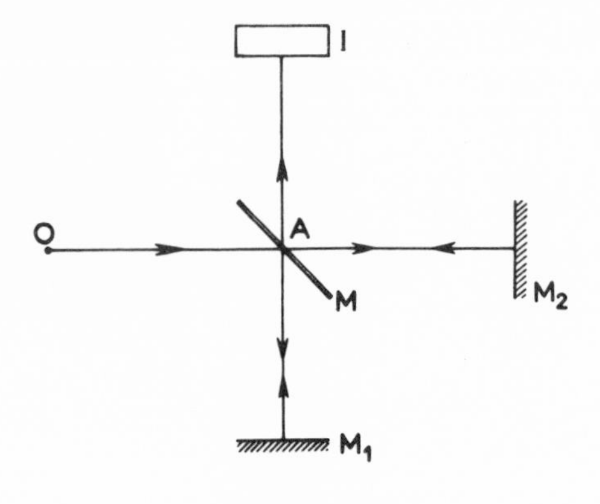

FIGURE 56

and the apparatus was designed in such a way that it could rotate around *A* and exchange the positions of the two rays. Any difference in the speed of light along the two paths should have been expressed by a change in the bands of interference, but this was not the case.* This demonstrates clearly that the speed of light is not added to the speed of the light source, any more than it is deducted from it. This particular speed always remains constant.

To account for this important fact, it is quite obvious that one must modify the first group of equations by which one passes from the coordinates of an event at reference *E* to the coordinates of this event at reference *E′*.

a2. THE TRANSFORMATION OF LORENTZ. The square of the speed of light is equal, on the one hand, to $\left(\frac{dx}{dt}\right)^2 + \left(\frac{dy}{dt}\right)^2 + \left(\frac{dz}{dt}\right)^2$, when x, y, and z are the coordinates of a photon (luminous particle) in the system of axes $Oxyz$, and on the other hand, it is equal to the square of the constant c^2. Therefore, for a photon,

$$c^2\overline{dt}^2 - \overline{dx}^2 - \overline{dy}^2 - \overline{dz}^2 = 0$$

* The device made it possible to detect a difference in speed of 1.5 kilometers per second, whereas the difference the experimenters expected to find was 30 kilometers per second.

But at reference E′ one will also have, since the speed of light there must be equal to c,

$$c^2\overline{dt'^2} - \overline{dx'^2} - \overline{dy'^2} - \overline{dz'^2} = 0$$

With the result that, for any particle at all, one will always have, when one changes references,

$$c^2\overline{dt'^2} - \overline{dx'^2} - \overline{dy'^2} - \overline{dz'^2} = f(x, y, z, t)$$
$$(c^2\overline{dt^2} - \overline{dx^2} - \overline{dy^2} - \overline{dz^2})$$

$f(x, y, z, t)$ being a certain function of x, y, z, t. One must find formulas for a changing reference which satisfy the foregoing equation. It can be demonstrated that $f(x, y, z, t)$ must be identical to 1, so that:

$$c^2\overline{dt'^2} - \overline{dx'^2} - \overline{dy'^2} - \overline{dz'^2} = c^2\overline{dt^2} - \overline{dx^2} - \overline{dy^2} - \overline{dz^2}$$

or in other words, the quantity

$$c^2\overline{dt^2} - \overline{dx^2} - \overline{dy^2} - \overline{dz^2}$$

is a constant in the transformation. It plays the role of the quantity $\overline{dx^2} + \overline{dy^2} + \overline{dz^2}$, which is a constant in a changing of purely spatial coordinates, and we shall assume that:

$$\overline{ds^2} = c^2\overline{dt^2} - \overline{dx^2} - \overline{dy^2} - \overline{dz^2}$$

By arranging the axes as above, we shall have $y' = y$ and $z' = z$, and one need only find a transformation of x and t into x' and t' which leaves the quantity $c^2\overline{dt^2} - \overline{dx^2}$ constant. Without making the relatively simple calculations that lead to the result, we shall confine ourselves to giving the formulas at which one arrives:

$$x' = \frac{x - vt}{\sqrt{1 - \frac{v^2}{c^2}}}; \quad y' = y; \quad z' = z; \quad t' = \frac{t - \frac{v}{c^2}x}{\sqrt{1 - \frac{v^2}{c^2}}} \qquad (2)$$

a3. SPACE-TIME IN SPECIAL RELATIVITY. Given the form of the second group of equations, it is no longer possible to assign a privileged position to the temporal coordinate of an event. We see here in effect that $t' = t$ provided v is zero, that is, when reference E' is at rest in relation to reference E. It is therefore quite natural to use as the basis of physics space-time references of four dimensions (since an event is located by four numbers x, y, z, and t) which are such that one passes from one of these references to a reference which is in rectilinear and uniform translation in relation to it by means of the formulas of Lorentz.

A reference of this kind is called a space-time of special relativity or a space-time of Minkowski. The distance between two neighboring events is the constant

$$\overline{ds}^2 = c^2\overline{dt}^2 - \overline{dx}^2 - \overline{dy}^2 - \overline{dz}^2 = c^2\overline{dt}^2 - \overline{d\sigma}^2$$

We shall designate as $\overline{d\sigma}^2$ the quantity $\overline{dx}^2 + \overline{dy}^2 + \overline{dz}^2$.

Although time is related to the coordinates of space, it is nevertheless true that it does not figure in the $\overline{ds}^2$ in the same way as these coordinates. Time in relativity, contrary to widespread opinion, retains a character of its own and does not play the same role as space.

The speed of a particle has as its square $\overline{d\sigma}^2/\overline{dt}^2$.

Since $\dfrac{\overline{ds}^2}{\overline{dt}^2} = c^2 - \dfrac{\overline{d\sigma}^2}{\overline{dt}^2}$, we deduce that the speeds of all particles are lower than c in absolute value, that is, lower than the speed of light, for the second member of the preceding equation must be equal to a square, and thus must be positive.

Now let us study the relative positions of two events x_1, y_1, z_1, t_1 and x_2, y_2, z_2, t_2 at reference E. (Let us call the events A_1 and A_2.) In the first place, is it possible to find a reference E' which is such that in this reference events A_1 and A_2 occur at the same point for an observer who is at rest in relation to this reference E'? If x'_1, y'_1 . . . ; x'_2, y'_2 . . . are the coordinates of A_1 and A_2 respectively in this new reference, it follows that

$$c^2(t_2 - t_1)^2 - (x_2 - x_1)^2 - (y_2 - y_1)^2 - (z_2 - z_2)^2 =$$
$$c^2(t_2 - t_1)^2 - \sigma^2 = c^2(t'_2 - t'_1)^2 - \sigma'^2$$

with $\sigma'^2 = (x'_2 - x'_1)^2 + (y'_2 - y'_1)^2 + (z'_2 - z'_1)^2$

We want σ' to be zero; consequently

$$s^2 = c^2(t_2 - t_1)^2 - \sigma^2 = c^2(t'_2 - t'_1)^2$$

s^2 is positive, therefore σ is lower in absolute value than $c\ (t_2 - t_1)$. Therefore, in order for the problem to be possible, it is necessary that the distance (in spatial terms) between the two events be shorter than the distance traveled by light during the interval of time $t_2 - t_1$ that separates the two events. If this is indeed the case, we say that the interval s between the two events is time-like. If events A_1 and A_2 involve the same particle P which has moved from P_1 to P_2 between two moments t_1 and t_2, the interval between A_1 and A_2 is necessarily a time interval, since the speed of the particle is necessarily lower than c. Let us also note that the time elapsed at reference E' between the two events is, according to the last formula arrived at,

$$t'_2 - t'_1 = \frac{1}{c}\sqrt{c^2(t_2 - t_1)^2 - \sigma^2}$$

It is not equal to the time that has elapsed at reference E, which is $t_2 - t_1$. This result, which is altogether logical in terms of the properties of space-time utilized, makes the theory of special relativity difficult for many people to understand.

Actually, an experimental proof of the reality of this phenomenon has been given by mesons. The mu meson is an unstable elementary particle which lasts only around 2×10^{-6} seconds when it is at rest. Mesons have been detected in cosmic rays which have traveled several kilometers through the atmosphere. Although these mu mesons have a speed close to the speed of light, their very short life duration would permit them to travel only around 600 meters. In fact, the length of the life of the mu meson is much greater for the terrestrial observer, in relation to whom the meson moves at a high speed, which permits it to cover a distance of around 6,000 meters before disintegrating.

Now let us see whether it is possible to find a reference E' such that an observer at rest in relation to it will observe that events A_1 and A_2 take place there simultaneously. It is still true that

$$s^2 = c^2(t_2 - t_1)^2 - \sigma^2 = c^2(t'_2 - t'_1)^2 - \sigma_2'^2$$

and this time, we want $t'_2 - t'_1$ to be zero; therefore

$$\sigma^2 = c^2(t_2 - t_1)^2 + \sigma'^2$$

The distance between the two events at E is greater than the distance traveled by light during interval $t_2 - t_1$. We say that interval s is spacelike.

To compare two events in a clearer way, let us suppose that these events take place along the x axis and that event A_1 is taken as the origin. Hence,

$$x_1 = y_1 = z_1 = t_1 = 0; \quad y_2 = z_2 = 0$$

We shall designate x_2 by x and t_2 by t. The interval between the two events is given by

$$s^2 = c^2t^2 - x^2$$

Let us make a diagram, starting with two axes with rectangular coordinates Ox and Ot (Fig. 57). An event is represented by

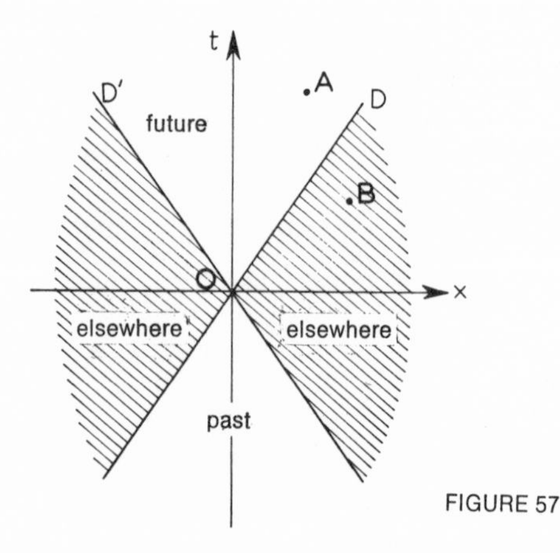

FIGURE 57

one point on the plane, the first event being represented by the origin O. Let A be the point representing the second event. A straight line passing through the origin represents the movement of a particle that is found at point $x = 0$ at time $t = 0$, at least in a case where this movement is possible. Lines D and D', whose equations are $x = ct$ and $x = -ct$, represent the movements of a luminous particle along the x axis, either in one direction or in the other. If the interval between events O and A is a time interval, point A is in the non-crosshatched part of the drawing, for $s^2 = c^2t^2 - x^2$ is positive. If, in addition, A is above Ox, t is positive, event A is subsequent to event O, and the corresponding zone of the diagram represents the future of O, that is, those points that can physically be reached by starting at O. The non-crosshatched area below Ox is, on the other hand, the past of O, that is, the points from which one can physically reach point O. The cross-hatched area corresponds to events which are such that the interval between these events and O is space-like. As we have seen, no physical action can connect O to such events; this is why this area is designated as "elsewhere."

We know that if interval OA is time-like, at a reference E' which is different from E, the time elapsed is

$$t' = \frac{1}{c}\sqrt{c^2t^2 - x^2}$$

t' is never zero (except if x and t are zero or if $ct = x$, a case peculiar to light). Therefore there is no reference for which O and A would be simultaneous, and consequently there is also no reference E' for which, if A is in the future of O at E, A is in the past of O at E'. The same is not true for a point B located in the area marked "elsewhere," but since there is no physical action possible between O and the points located in its "elsewhere," the principle of the anteriority of cause to effect is respected in special relativity.

Now let us see what becomes of the law of the composition of speeds. Let us suppose that a particle is moving along axis Ox at reference E with a uniform speed V, and let V' be its speed at E' along axis $O'x'$.

We have: $V = \frac{dx}{dt}$; $V' = \frac{dx'}{dt'}\cdot$ Now, the formulas of Lorentz give us, after differentiation:

$$\frac{dx'}{dt} = \frac{V - v}{\sqrt{1 - \frac{v^2}{c^2}}}; \quad \frac{dx'}{dt'} = V' = \frac{dx'}{dt}\frac{dt}{dt'}; \quad \frac{dt}{dt'} = \frac{\sqrt{1 - \frac{v^2}{c^2}}}{1 - \frac{vV}{c^2}}$$

Therefore

$$V' = \frac{V - v}{1 - \frac{vV}{c^2}}$$

$$V = \frac{V' + v}{1 + \frac{vV'}{c^2}}$$

If v/c is small enough to be disregarded, we have $V = V' + v$; moreover, in this case, the formulas of Lorentz give us

$$x' = x - vt, \quad y' = y, \quad z' = z, \quad t' = t$$

and we encounter the usual formulas, which are adequate when the references chosen have, in relation to one another, speeds that are low in comparison with the speed of light. We also note, in the formula governing the composition of speeds, that no matter what the value of v is, if $V' = c$, $V = c$.

a4. DYNAMICS OF SPECIAL RELATIVITY. In Newtonian mechanics, the equation for the motion of a particle is written

$$m\vec{\gamma} = \vec{F}$$

m being the mass of the particle, $\vec{\gamma} = \frac{d\vec{v}}{dt}$ being the acceleration vector, the derivative with respect to time of the velocity vector $\vec{v}$, and $\vec{F}$ being the vector force applied to the particle. Since v is velocity, the following equivalence may be deduced from the preceding equation:

$$\frac{d}{dt}(\tfrac{1}{2}mv^2) = \vec{F} \cdot \vec{v}$$

which expresses the fact that the derivative in relation to time of "kinetic energy," $\tfrac{1}{2}mv^2$, is equal to the scalar product of vectors $\vec{F}$ and $\vec{v}$. (The scalar product of two vectors is equal to the product of the lengths of the two vectors times the cosine of the angle between their directions.)

In special relativity, one is obliged to replace the above equivalence by the equation

$$\frac{d}{dt}\left(\frac{m_0c^2}{\sqrt{1-\frac{v^2}{c^2}}}\right) = \vec{F} \cdot \vec{v}$$

The quantity

$$E = \frac{m_0c^2}{\sqrt{1-\frac{v^2}{c^2}}}$$

will be known as total energy, by analogy with kinetic energy in Newtonian mechanics; m_0 is a constant called the "rest mass" of the particle. If v is zero, $E = m_0c^2$. If v is not zero, but is small in relation to c, one can write

$$\frac{1}{\sqrt{1-\frac{v^2}{c^2}}} \sim 1 + \frac{1}{2}\frac{v^2}{c^2}$$

for the second member of this equivalence is only slightly different from the first. Therefore,

$$E = m_0c^2 + \frac{1}{2}m_0v^2$$

We see that total energy is equal to energy at rest plus kinetic energy.

The fundamental equation of mechanics in Newtonian mechanics was written

$$\frac{d}{dt}(m\vec{v}) = \vec{F}$$

In special relativity, it will be written

$$\frac{d}{dt}\left(\frac{m_0\vec{v}}{\sqrt{1-\frac{v^2}{c^2}}}\right) = \vec{F}$$

It is analogous to the preceding equation, provided we regard the particle as having a mass that varies with velocity

$$m = \frac{m_0}{\sqrt{1-\frac{v^2}{c^2}}}$$

Mass m is equal to the rest mass if v is zero; it is infinite if v is equal to the speed of light.

Finally, seeing that

$$E = \frac{m_0c^2}{\sqrt{1-\frac{v^2}{c^2}}}$$

we deduce the equation

$$E = mc^2$$

This equation, which reveals the equivalence between mass and energy, plays a fundamental role in modern physics.

It should be noted that the fact that mass increases with

velocity, expressed by the relationship given above, is true for an observer connected to a reference (E) in relation to which the particle in question has a motion of rectilinear and uniform translation of velocity v. In a reference connected to the particle, the latter has zero velocity and the mass of the particle is its rest mass; it is, on the contrary, the mass of the observer that is multiplied by the inverse of $\sqrt{1 - v^2/c^2}$.

b. RIEMANNIAN GEOMETRY

b1. MANIFOLDS WITH N DIMENSIONS. To determine the position of a point on a plane, we need to know two numbers; for example, the abscissa x and the ordinate y of the point. Moreover, it is possible, on a plane, to define in a precise way what are referred to as "neighboring points," and in addition, each point P on the plane has an infinity of neighboring points—all points situated within a circle whose center is P and whose radius is arbitrarily small. We say that the plane is a manifold of two dimensions. Since we can define on the plane functions and their derivatives, we say that the manifold is differentiable, but we shall not insist on this point.

The surface of a sphere is also a manifold of two dimensions. One must have two numbers in order to locate a point on it without ambiguity (for example, longitude and latitude), and one can also define without ambiguity the vicinity of a point on it. In a space of ordinary geometry, however, each point is located by three numbers (for example, the three coordinates x, y, and z in relation to axes Ox, Oy, and Oz); this space is a manifold of three dimensions. One can broaden this to include manifolds with n dimensions, n being any whole number. We have studied the space-time of special relativity; it is a manifold of four dimensions: the three coordinates of space, and time.

b2. $\overline{ds}^2$; EUCLIDEAN SPACES. Let us consider a manifold of two dimensions for the sake of simplicity. Let u and v be the coordinates of a point (these do not necessarily have to be rectangular coordinates, but two parameters which permit us to locate the point) and let $u + du$ and $v + dv$ be the coordinates of a neighboring point; du and dv are small. Now let us consider a function ds defined by its square in the following manner:

$$\overline{ds^2} = g_{11}(u,v)\ \overline{du^2} + g_{12}(u,v)\ du\ dv + g_{22}(u,v)\ \overline{dv^2}$$

g_{11}, g_{12}, and g_{22} are given functions of u and v. We shall say that we have chosen a $\overline{ds^2}$ (read "*d*, *s*, squared") on the manifold, and we shall agree to call the quantity ds the distance between two neighboring points. If two points whose coordinates are u_1, v_1 and u_2, v_2 are not neighboring points, we shall calculate the distance between these two points by finding the integral of ds between the two points along a line C.

For example, let us define the $\overline{ds^2}$ on a manifold of two dimensions by the formula

$$\overline{ds^2} = \overline{du^2} + u^2\overline{dv^2}$$

Here

$$g_{11} = 1, \quad g_{12} = 0, \quad g_{22} = u^2$$

One can change systems of coordinates and locate a point P whose coordinates are u and v by means of two other parameters, x and y, for example, connected to u and v by certain relationships. One can then calculate the new form taken by the $\overline{ds^2}$ by calculating the increments du and dv and the functions g_{11}, g_{12}, etc., in terms of the increments (called differential increments) dx, dy of x and y.

For example, in the example given above, let us assume that

$$x = u \cos v, \quad y = u \sin v$$

or

$$u = \sqrt{x^2 + y^2}, \quad \tan v = \frac{y}{x}$$

We shall find

$$\overline{ds^2} = \overline{dx^2} + \overline{dy^2}$$

By using these new parameters x and y, the $\overline{ds^2}$ takes on a very simple form; indeed, the coefficients of $\overline{dx^2}$ and $\overline{dy}^2$ are equal to one, and that of the product $dx\,dy$ is zero. In the general formula given above, we would have here $g_{11} = g_{22} = 1$, $g_{12} = 0$. This is true in our example, for any point of coordinates u and v in the manifold: we say that the manifold on which we have defined such a $\overline{ds^2}$ constitutes a Euclidean space of two dimensions.

Generally speaking, we shall say that a manifold of n dimensions constitutes a Euclidean space of n dimensions if one has defined on this manifold a $\overline{ds^2}$ which is such that, by an appropriate change of variables, one can, *on any point of the manifold*, reduce the $\overline{ds^2}$ to the following form: the coefficients of the squares of the differentials are equal to $+1$ or -1, the coefficients of the products two at a time of the differentials are zero.

Thus, the space-time of special relativity is a Euclidean space of four dimensions. In effect,

$$\overline{ds^2} = c^2\overline{dt^2} - \overline{dx^2} - \overline{dy^2} - \overline{dz^2}$$

By assuming that $ct = T$, we have $cdt = dT$, and

$$\overline{ds^2} = \overline{dT^2} - \overline{dx^2} - \overline{dy^2} - \overline{dz^2}$$

In the space of two dimensions studied above, we had $\overline{ds^2} = \overline{dx^2} + \overline{dy}^2$, the coefficients of the squares of the differentials were $+1$ and $+1$; we say that the signature of the $\overline{ds^2}$ is $+\,+$. In a space of special relativity, we see that the signature is $+\,-\,-\,-$. Such a Euclidean space, for which one does not have only pluses in the signature, is called pseudo-Euclidean, in distinction to true Euclidean spaces.

b3. RIEMANNIAN SPACES; GEODESICS. Let us consider a Riemannian manifold of n dimensions. A point P is located on this manifold by n parameters x^1, x^2, etc. x^n or x^α $(\alpha = 1, 2, \ldots n)$. The second notion signifies that one must give the index α all the values 1, 2, . . . n in succession in order to have the n coordinates of point P. If $x^\alpha + dx^\alpha$ ($\alpha = 1, 2, \ldots n$) designate the coordinates of a point in the vicinity of P, we shall define on the manifold a $\overline{ds^2}$ at each point by the formula

$$\overline{ds^2} = g_{\alpha,\beta}\, dx^\alpha dx^\beta (\alpha \text{ and } \beta = 1, 2, \ldots n)$$

The second member of the above formula must be understood in the following manner: α and β take on respectively all the values 1, 2, . . . n and one adds up all the terms found. For example, if $n = 2$ (the manifold has two dimensions), we will have

$$\overline{ds^2} = g_{11}dx^1\, dx^1 + g_{12}dx^1dx^2 + g_{21}dx^2dx^1 + g_{22}dx^2dx^2$$

(When we write dx^2 instead of $\overline{dx^2}$, we are referring, not to the square of dx, but to the differential of x^2, 2 being here the superscript, or number, of the coordinate.) The quantities $g_{a\beta}$ are functions of the coordinates $x^1, x^2, \ldots x^n$ and we assume that $g_{a\beta} = g_{\beta a}$. One can arrange the $g_{a\beta}$'s in a table with n lines and n columns in the following manner:

$$\begin{matrix} g_{11} & g_{12} & g_{13} & \cdots & g_{1n} \\ g_{21} & g_{22} & g_{23} & \cdots & g_{2n} \\ \cdot & \cdot & \cdot & \cdots & \cdot \\ g_{n1} & g_{n2} & g_{n3} & \cdots & g_{nn} \end{matrix}$$

These coefficients $g_{a\beta}$ are called the coefficients of the metric of the manifold defined by knowledge of the $\overline{ds^2}$. (We assume that the determinant of the above table is not zero.)

The $g_{a\beta}$'s constitute the twice covariant components of a mathematical entity called a "tensor of the second order," which in this case is called a metric tensor.

Now let us suppose that it is impossible to find a change of variables by which we may pass from variables x^a to variables u^a in such a way that at every point of the manifold, the $\overline{ds^2}$ takes the form of the $\overline{ds^2}$ of a Euclidean space: in this case we say that we are dealing with a Riemannian space. The manifold with n

dimensions on which we will have defined such a $\overline{ds}^2$ will constitute a Riemannian space of n dimensions. We shall see a simple example of a Riemannian space further on.

Generally speaking, we shall be considering Riemannian spaces in which all the $g_{\alpha\beta}$'s are zero if α is different from β. In other words, in the table given above, all the elements are zero except those situated on the diagonal running from the upper left corner to the lower right corner. The + or − signs found in front of the remaining $g_{\alpha\beta}$'s thus constitute the signature of the metric.

Now let us consider a given point in the manifold, P_0 whose coordinates are $x_0{}^\alpha$ ($\alpha = 1, 2, \ldots n$); if we replace the x_α's by $x_0{}^\alpha$'s in the $g_{\alpha\beta}$'s, the $\overline{ds}^2$ becomes

$$\overline{ds}^2 = g^0{}_{\alpha\beta}\, dx^\alpha dx^\beta$$

The coefficients are constant and one will always be able to find variables such that this $\overline{ds}^2$ takes the form of a $\overline{ds}^2$ of a Euclidean space. For example, if $g_{\alpha\beta} = 0$, when α is different from β, we have

$$\overline{ds}^2 = g^0{}_{\alpha\alpha}\overline{dx}^{\alpha 2}$$

and we shall take as a variable

$$\sqrt{|g^0{}_{\alpha\alpha}|}\, x^\alpha = u^\alpha$$

A Euclidean space that has a $\overline{ds}^2$ of this kind is called the Euclidean space tangent to the Riemannian space at point P_0. It is considered that to a certain extent this space coincides with the Riemannian space in a vicinity of P_0.

We shall now study some curves which play a very important role in the theories with which we are concerned here.

If the coordinates x^α of a point P are not independent, but are all functions of a single parameter μ, then point P is not located at random on the manifold. It is located on a curve whose equations are defined by

$$x^\alpha = f_\alpha(\mu),\ \alpha = 1, 2, \ldots n$$

If two points P and Q are on the same curve, the distance between the two points on the curve will be obtained by calculating the integral of ds taken along the curve between P and Q.

Now, it can be demonstrated that there exist certain curves C which are such that the distance between two points P and Q along curve C which passes through these two points is the shortest possible distance.*

Curves of this kind are called geodesics of space. It is interesting to note that these curves depend solely on the $g_{\alpha\beta}$'s. If the metric tensor is given, all the geodesics are completely determined.

In a Euclidean space, geodesics are called straight lines. Thus one encounters the well-known property that in a plane (i.e., a Euclidean space of two dimensions), a straight line is the shortest distance between two points. In a space-time of special relativity (Minkowski's space), geodesics are the "straight lines" of this space-time, that is, the straight lines of space traveled by uniform motion. If, in addition, we are dealing with straight lines traveled by light,

$$\overline{ds}^2 = c^2\overline{dt}^2 - \overline{dx}^2 - \overline{dy}^2 - \overline{dz}^2 = 0$$

the distance between any two points on such a geodesic is zero. Thus in Minkowski's space, light travels along geodesics of zero dimensions. (In the plane of ordinary geometry, there are no geodesics of zero dimensions, for this space is, properly speaking, Euclidean.)

Now let us look at certain properties of geodesics from the point of view of mechanics. In classical Newtonian mechanics, we know that if the motion of a particle is related to an absolute reference and if the particle is not subjected to any force, it describes a straight line by uniform motion; consequently, it describes a geodesic of Euclidean space in relation to which one defines the motion. This property is retained if one relates the motion of the particle to another reference of inertia. This is likewise the case in special relativity. One need only refer to the equations we have studied above.

* Or the longest possible distance.

Still placing ourselves within the context of Newtonian mechanics, we can demonstrate that a particle which is obliged to move in a Riemannian space and which is not subjected to any force describes a geodesic of this space. When we say that the particle is not subjected to any force, we mean no force other than the force of restraint, which expresses the fact that the particle is obliged to move in a particular space.

Thus, a nonplanar surface constitutes a Riemannian space. A material point obliged to remain on the surface is subject to a binding force called the surface reaction, and possibly, to an additional force. If this additional force is zero, the point describes a geodesic of the surface.*

b4. THE CURVATURE OF RIEMANNIAN SPACES. Before trying to define what is meant by the curvature of a Riemannian space, we shall give a particularly simple example of such a space (Fig. 58).

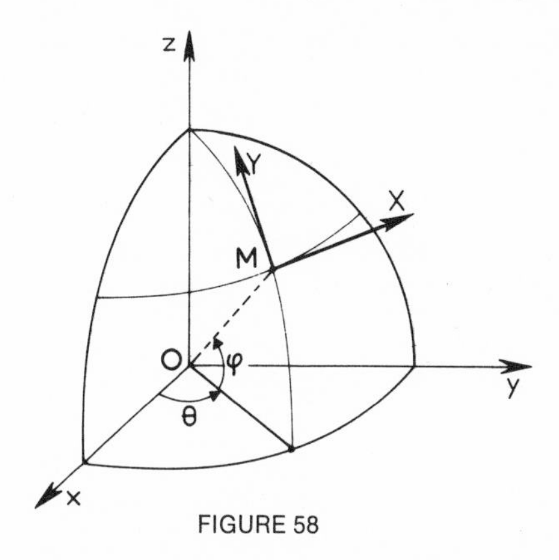

FIGURE 58

In the three-dimensional space of ordinary geometry, let us consider a sphere whose center is O and whose radius is 1. If we take a system of rectangular axes $Oxyz$ originating at O, the coordinates of a point M on the sphere are related by the equation

$$x^2 + y^2 + z^2 = 1$$

* Provided friction is zero.

which is the equation of the sphere and which expresses the fact that the square of OM is constant and is equal to 1. One can locate a point on the sphere M by its longitude θ and its latitude ϕ, and it will be true of M that

$$x = \cos\phi \cos\theta, \quad y = \cos\phi \sin\theta, \quad z = \sin\phi$$

We would find that

$$\overline{dx}^2 + \overline{dy}^2 + \overline{dz}^2 = \overline{d\phi}^2 + \cos^2\phi \, \overline{d\theta}^2$$

Any point M on the sphere can be located by giving two parameters θ and ϕ; the surface of the sphere is a manifold of two dimensions. Moreover, if we assume that

$$\overline{ds}^2 = \overline{d\phi}^2 + \cos^2\phi \, \overline{d\theta}^2$$

the surface of the sphere is a space characterized by a metric, but this space is not Euclidean. The shortest line joining two points on the sphere is not a straight line (indeed, there are no straight lines on the sphere), but an arc of a great circle passing through these two points. The geodesics of the sphere are the great circles of the sphere, they are not infinite and they all intersect. Contrary to the plane of ordinary geometry, the surface of the sphere is

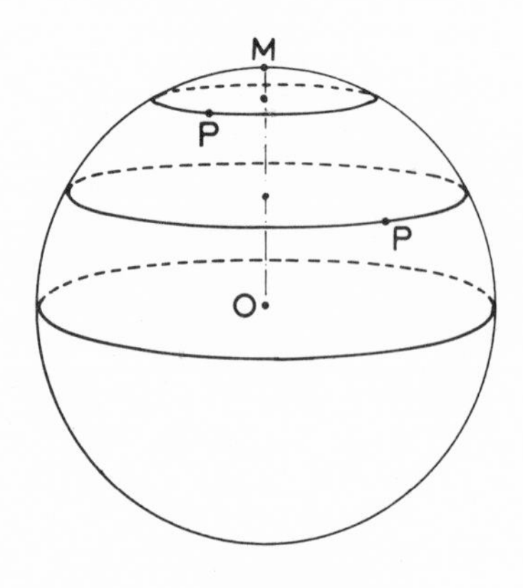

FIGURE 59

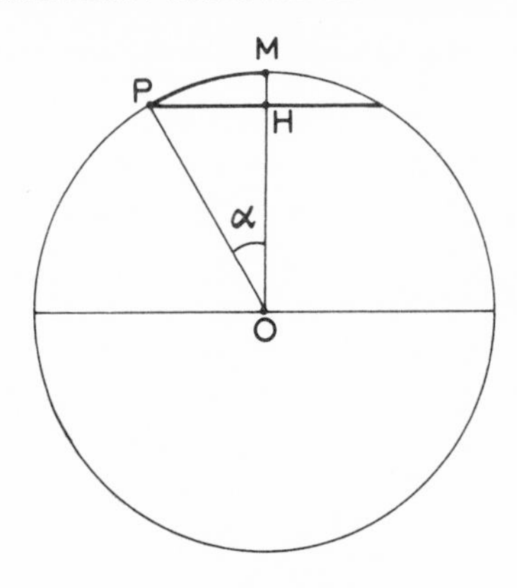

FIGURE 60

not a Euclidean space of two dimensions, but a Riemannian space of two dimensions.

It is easy to see that aside from the particular characteristics of geodesics, such a space has properties very different from those of the plane. In the first place, one cannot construct figures as large as one likes on it. For example (Fig. 59), let us rest one leg of a compass at *M* and the other at a neighboring point on the sphere. By rotating the compass we can draw a small circle whose center is *M* on the sphere and whose radius is *MP*. We can draw circles with larger radii whose centers are *M*, but the largest possible circle will be the great circle of the sphere whose pole is *M*, that is, the circle situated in the plane perpendicular to *OM* and passing through *O*. Moreover, in the plane, the ratio of the length of the circumference of a circle to its radius is equal to 2π. What is it on the sphere? The length of the circumference of radius *MP* is $L = 2\pi HP = 2\pi R \sin \alpha$, R being the radius of the sphere and α the angle *POM* (Fig. 60), and radius MP is equal to αR.

Therefore,

$$\frac{L}{MP} = 2\pi \frac{\sin \alpha}{\alpha}$$

If α is very small, $\sin \alpha$ is only slightly different from α and L/MP is very close to 2π; when α is not small, the ratio is lower than 2π. For $\alpha = \pi/2$, L/MP is equal to 4 (whereas $2\pi = 6.28...$).

Similarly, we would observe that the sum of the angles of a triangle is not equal to two right angles but that it is greater. This is obvious, especially for a triangle such as MAB in Fig. 61, which has two right angles A and B.

All the properties we have just seen express the fact that the surface of a sphere is a Riemannian space rather than a Euclidean space. Such a space possesses at every point a certain curvature that can be mathematically defined in terms of the coefficients $g_{\alpha\beta}$ of the $\overline{ds}^2$. The case of the sphere is simple, for this curvature is the same at every point. In any Riemannian space the curvature varies, in general, from one point to another. It may be positive (as in the case of the sphere) or negative. In the latter case, the sum of the angles of a triangle is less than two right angles, the ratio of the length of a circumference to its radius is greater than 2π, etc. If the curvature is zero, the space is Euclidean.

This notion of curvature may be extended to a Riemannian space with any number of dimensions; it is mathematically expressed by formulas which make use of the $g_{\alpha\beta}$'s and which we shall not give here but which are completely analogous to the formulas for the curvature of a two-dimensional surface in ordinary geometry. Naturally, in a three-dimensional space like our own it is impossible to "see" this curvature; only measurements

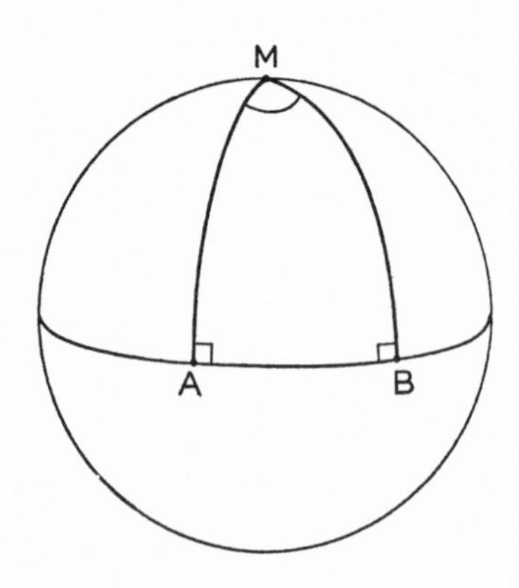

FIGURE 61

on figures of large dimensions enable us to conclude that this space is not Euclidean and to calculate its radius. This is what might be done by small two-dimensional creatures living on a large sphere which on first encounter they would believe to be flat. By measuring the lengths of very large circles, they would note a deviation from 2π in the ratio of their lengths to the radii. Furthermore, they might possibly observe that by following a geodesic (which for them is a "straight line"), they would return to the same point.

Let us return to Fig. 58. In a small area around *M*, it is possible to identify the surface of the sphere with the plane tangent to the sphere at this point. It will even be possible to relate this tangent plane to two axes of rectangular coordinates originating at *M*, *MX* and *MY*. For example, *MX* will be the tangent to the parallel of *M* and *MY* the tangent to the meridian. The Euclidean space of two dimensions related to axes *MX* and *MY* will be the Euclidean space tangent at *M* to the Riemannian space, as we have defined it in the preceding paragraph.

Let ϕ_0 and θ_0 be the coordinates of *M*. In the immediate vicinity of *M* we will have

$$\overline{ds}^2 = \overline{d\phi}^2 + \cos^2\phi_0\, \overline{d\theta}^2$$

and, if we let $\phi = Y$ and $\theta \cos \phi_0 = X$,

$$\overline{ds}^2 = \overline{dY}^2 + \overline{dX}^2$$

which is the $\overline{ds}^2$ of a Euclidean space.

c. THE THEORY OF GENERAL RELATIVITY

c1. PROPERTIES OF GRAVITATIONAL FIELDS. From the law of universal gravitation, on the one hand, and from the law $\vec{F} = m\vec{\gamma}$ on the other, it follows that the acceleration acquired by a body in a gravitational field is independent of its mass. To be more precise, let us imagine a body *M* of mass *m* and let us assume that it is subject to an attraction located at a point *O* due to a unit mass, with $OM = r$. We have

$$\vec{F} = \frac{Gm\,\overrightarrow{MO}}{r^3}$$

(G being the constant of universal gravitation); hence,

$$\vec{\gamma} = \frac{G\,\overrightarrow{MO}}{r^3}$$

We have assumed O to be fixed; it will be, for example, the origin of the system $Oxyz$ which forms a reference of inertia in which, consequently, the law $\vec{F} = m\vec{\gamma}$ is valid.

But the property which we have just observed for a point M subject to forces of gravitation is equally true if M is subject to forces of inertia. In effect, if the motion of a point M is no longer determined in relation to a system of inertia but in relation to a system of axes which does not have a rectilinear and uniform motion in relation to the reference of inertia, we have seen in the second part (IIIb3) that M is subject, in addition to the given forces, to an "imaginary" force of inertia $-m\vec{\gamma}_{OA}$. In this term $\vec{\gamma}_{OA}$ is the original acceleration of the noninertial reference. (We are assuming here that this reference is moving by translation; the forces of inertia would have to be expressed in a more complicated way if it were also moving by rotation.) It is clear that here again the acceleration acquired by M as a result of the force of inertia is independent of m.

This property is sometimes presented in an analogous but more intuitive form by saying that the gravitational mass of a body is equal to its inertial mass. Gravitational mass is the ratio between the force of gravity acting on a body and the acceleration acquired by this body when falling freely in a vacuum. Inertial mass is the ratio between a force of any kind other than gravity and the acceleration which it gives to a body.

This property is altogether remarkable; it is peculiar to gravitational fields and is not true for electric or magnetic fields.

The most important idea in the theory of general relativity is precisely that by considering this property, it ceases to regard the forces of gravitation as given forces of a particular kind, but brings them, together with the forces of inertia, back into the domain of forces caused by the reference being utilized to study the motion. We shall see that these forces will disappear as given forces, and that their effects will be regarded as properties of space-time.

It should be noted, however, that there is nevertheless a qualitative difference between the forces of inertia and the forces of gravitation. In the first place, one can cause forces of inertia to disappear by a timely selection of the reference to which one refers the motion; this is not true of forces of gravitation. Furthermore, at infinity, the gravitational field caused by a certain mass of matter becomes zero; on the contrary, the forces of inertia are not. For example, in a system rotating at a uniform rate of speed around an axis Oz, the force of inertia (called centrifugal force) becomes infinite when one moves an infinite distance away from axis Oz. Thus one will try to carry out the proposed aim in a relatively limited area in which the gravitational field will be regarded as constant. The theory of relativity will be a local theory of gravitation.

c2. GEOMETRY OF SPACE RELATED TO A NON-GALILEAN REFERENCE. Let us consider a system of Cartesian coordinates related to a Galilean reference (Fig. 62). As we have seen, its $\overline{ds^2}$ is given in special relativity by

$$\overline{ds^2} = c^2\overline{dt^2} - \overline{dx^2} - \overline{dy^2} - \overline{dz^2}$$

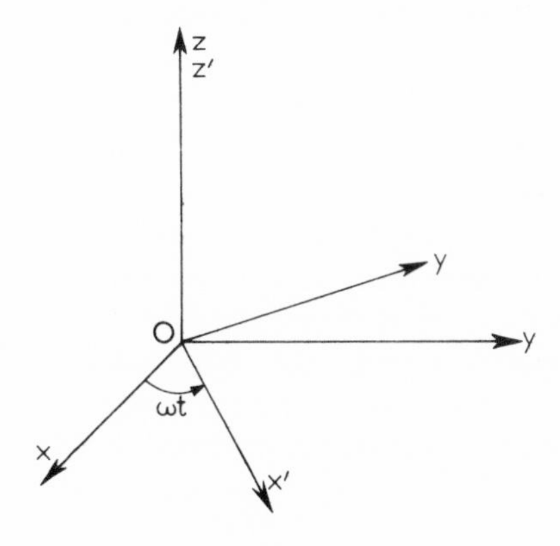

FIGURE 62

Now let us suppose that there is another system $Ox'y'z'$ which is not Galilean. Let us suppose, for example, that Oz' coincides with Oz and that system $Ox'y'z'$ is rotating at a uniform rate ω around Oz and that at instant $t = 0$, Ox' coincides with Ox. We have

$$\begin{aligned} x &= x' \cos \omega t - y' \sin \omega t \\ y &= x' \sin \omega t + y' \cos \omega t \\ z &= z' \end{aligned}$$

The $\overline{ds}^2$ will become

$$\overline{ds}^2 = [c^2 - \omega^2(x'^2 - y'^2)]\overline{dt}^2 - \overline{dx}'^2 - \overline{dy}'^2 - \overline{dz}'^2 + 2\omega y' dx' dt - 2\omega x' dy' dt$$

No matter what expression one uses for time t' in reference $Ox'y'z'$, one will not be able to reduce the above formula to the formula for a Euclidean $\overline{ds}^2$. In other words, in a reference of this kind, the $\overline{ds}^2$ will take the form:

$$\overline{ds}^2 = g_{\alpha\beta} dx^\alpha \, dx^\beta (\alpha, \beta = 1,2,3,4)$$

and no change in the variables will enable one to arrive at a Euclidean space. The space will be Riemannian, and, in particular, it will have a curvature.

Let us look somewhat more closely at this example, which is very classical but very instructive. Let us suppose that in place $Ox'Oy'$ there is a circular platform of radius a and of center O that rotates around axis Oz, which is perpendicular to it, at a uniform rate ω (Fig. 63).

An observer measures the length of a radius OA. To do this, he places end to end small identical rules the length of OA and finds a dimension a. To measure the circumference of the platform, he places the rules perpendicularly to the radius. If the rule has a length l in relation to the fixed reference $Oxyz$ and if the measurement is made within a short interval during which one can regard the rule as having a rectilinear and uniform motion

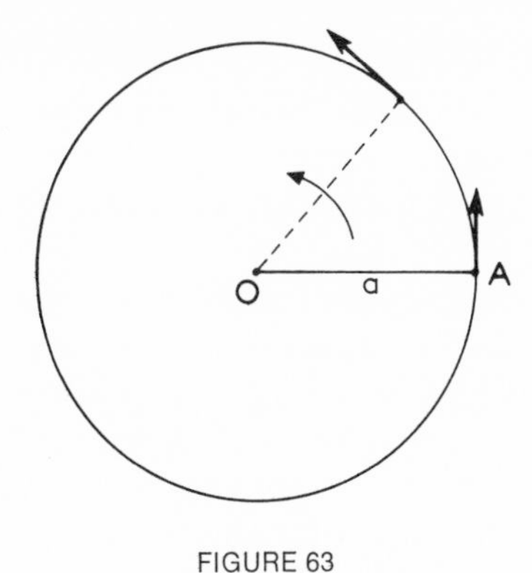

FIGURE 63

whose velocity $v = a\omega^2$, for an observer bound to the platform, the length of this rule will be:

$$l' = l\sqrt{1 - v^2/c^2}$$

according to the formulas of Lorentz.

Hence for the observer, the length of the circumference will be greater than $2\pi a$. In other words, the ratio of the length of the circumference to the radius will be greater than 2π, and from this the observer will deduce that his space is not Euclidean and that it has a negative curvature.

c3. THE $\overline{ds}^2$ OF GENERAL RELATIVITY. By combining what we have seen in the two preceding paragraphs, we see how general relativity interprets the phenomenon of gravitation. The presence of masses at a point in space distorts this space by changing its geometry, which is no longer that of a Euclidean space, but is now that of a Riemannian space. The presence of matter curves the space in its vicinity.

In particular, light does not describe straight lines, but geodesics of curved space-time, with the result that in the vicinity of masses, the paths of light rays must be curved inward. This has been observed during eclipses of the sun; stars which are very close to the edge of the sun at the time of the eclipse seem to have shifted with respect to the positions they would normally occupy. This is caused by the fact that the light rays coming from these stars are passing very close to the edge of the sun at the time of observation and that the curvature of space-time is there very great.

Another consequence of the theory of general relativity which has lent itself to experimental confirmation is the shift toward red of spectral rays when the atoms that emit them find themselves in an intense field of gravitation. In the case of the sun, this shift is only 0.011 angstroms, but for the white dwarfs, which are extremely dense stars, the shift is much greater and can be perfectly observed. This effect has also been demonstrated on earth in the course of an experiment conducted at Harvard: γ rays were shifted toward red by terrestrial gravitation.

The problem that arises, then, is the following: given the distribution of matter at a point in space, how is one to determine the geometry of the space in its immediate vicinity? In other words, how is one to determine the $\overline{ds}^2$ of this space, that is, the $g_{\alpha\beta}$ coefficients? To do this one must try to extend the properties of the universe of special relativity in which free particles describe geodesics and light describes geodesics of zero length. In the space-time of general relativity the material points which are subject only to the force of local gravitation will describe the geodesics of this space, with the result that the forces of gravitation disappear as given forces and become, in some sense, comparable to the forces of restraint of a material point bound to a surface.

Einstein's equations generalize the equations of Laplace and Poisson for determining the potential locally; they also establish relationships between tensors in a space-time of four dimensions. From these equations, one can deduce certain $\overline{ds}^2$'s in accordance with the given initial conditions. We shall return to this point later; first, we shall briefly study one particular $\overline{ds}^2$.

c4. SCHWARZSCHILD'S $\overline{ds}^2$. This $\overline{ds}^2$ is a solution to Einstein's equations in the simple case in which there is a central distribution of matter, that is, when the $\overline{ds}^2$ is the same for all points equidistant from a center. Furthermore, one is located in a vacuum, that is, beyond any masses which create a field of gravitation. We have, therefore,

$$\overline{ds^2} = \left(c^2 - \frac{2\mu}{r}\right)\overline{dt^2} - \frac{\overline{dr^2}}{1 - \frac{2\mu}{c^2 r}} - r^2\left(\overline{d\phi^2} + \cos^2\phi\,\overline{d\theta^2}\right)$$

In this formula, r, ϕ, and θ are coordinates analogous to the ordinary spherical coordinates (distance of the particle from point O, latitude and longitude), c is the speed of light, and μ is proportional to the central mass situated at O.

If μ is zero, that is, if there is no mass, the $\overline{ds^2}$ becomes

$$\overline{ds^2} = c^2\overline{dt^2} - \overline{dr^2} - r^2\left(\overline{d\phi^2} + \cos^2\phi\,\overline{d\theta^2}\right)$$

By assuming that

$$x = r\cos\phi\cos\theta, \quad y = r\cos\phi\sin\theta, \quad z = r\sin\phi$$

the $\overline{ds^2}$ is reduced to the following form:

$$\overline{ds^2} = c^2\overline{dt^2} - \overline{dx^2} - \overline{dy^2} - \overline{dz^2}$$

The space is therefore Euclidean. The same is true, moreover, if μ is not zero and r is large, that is, if one is at a distance from the masses.

Let us suppose that the mass situated at O is the mass of the sun and that one is studying the motion of a small mass, for example, the planet Mercury, located in its vicinity. One will look for the geodesics of Schwarzschild's $\overline{ds^2}$. The calculation is not very difficult. One finds that the motion occurs in a plane passing through O. The path is very close to an ellipse of focus O described according to the law of areas, but its perihelion is not fixed and moves forward, in the case of Mercury, by 42.9 seconds per century. This corresponds exactly to the hitherto unexplained discrepancy between the theoretical advance caused by the classic perturbations and the advance provided by observation.

d. THE STRUCTURE OF MATTER; THE STARS

d1. FUNDAMENTAL PARTICLES; THE ATOM; THE MOLECULE. The principal fundamental particles are the proton, the neutron, and the electron. These particles are characterized by their masses and their electrical charges. The proton has a positive elementary charge, the electron has an equal but negative charge; on the other hand, the mass of the electron is 1,836 times smaller than that of the proton. The neutron has a mass almost equal to that of the proton, but it is electrically neutral, which means that its charge is zero.

Protons and neutrons are combined to form nuclei; hence, protons and neutrons are designated by the term "nucleons." The electrical charge of the nucleus is positive; it is equal to as many times the electrical charge of the proton as there are protons in the nucleus. There exist throughout nature 325 distinct types of nuclei, of which 274 are stable and 51 are unstable.

A nucleus is characterized by two numbers: the atomic number Z, which is the number of protons it contains, and the "mass number" A, which is the total number of nucleons in the nucleus. Consequently, a nucleus whose atomic number is Z and whose mass number is A contains $A - Z$ neutrons.

A nucleus has a positive electrical charge equal to Z times the elementary electrical charge of the proton; it may be surrounded by Z electrons, which gravitate around it. The whole unit is then electrically neutral and constitutes a neutral atom. If the number of electrons that surrounds the nucleus is lower than Z, the atom has a positive electrical charge; we say that it is ionized. An atom is said to be singly ionized if its nucleus is surrounded by $Z - 1$ electrons, doubly ionized if it is surrounded by $Z - 2$ electrons, etc. An atom that is completely ionized is an atom reduced to its nucleus.

Two atoms whose nuclei contain the same number of protons but a different number of neutrons are regarded as two atoms of the same element. They have similar chemical properties, but naturally their masses are different. These atoms are given the same name and they are represented by the same letter to which will be assigned, if necessary, a superscript equal to the mass number. Two atoms of the same element which differ in their number of neutrons are known as two isotopes of the same element.

Let us take a few examples. The simplest atom is the hydrogen atom represented by the letter H; it contains one proton and one electron. Therefore, $Z = 1$, and $A = 1$, and one can represent this atom either by H, or by $_1H^1$ (Z being a subscript, *A* being a superscript), or simply by H^1, for the mere fact of writing H signifies that we are dealing with hydrogen and therefore that $Z = 1$, and it is unnecessary to state this fact. There exists an isotope of hydrogen, $_1H^2$ (or H^2), whose nucleus contains one proton and one neutron; it is sometimes designated by the name deuterium.

The element oxygen (O) has three isotopes: $_8O^{16}$, $_8O^{17}$, $_8O^{18}$; the first has 8 protons and 8 neutrons, the second 8 protons and 9 neutrons, and the third 8 protons and 10 neutrons. In all cases, the number of protons is 8, which is the atomic number of oxygen; naturally, the corresponding neutral atoms have 8 electrons.

In the presence of a sample of any element, one finds that the proportion of atoms of the different isotopes of this element is absolutely constant. Often one of them clearly predominates; thus, a certain mass of hydrogen contains 99.985 percent of H^1 and 0.015 percent of H^2; a certain mass of oxygen contains 99.759 percent of O^{16}, 0.037 percent of O^{17}, and 0.204 percent of O^{18}; sometimes two isotopes are found in close proportions (silver contains 51.9 percent of Ag^{107} and 48.1 percent of Ag^{109}).

Another very important atom is helium. Its atomic number, Z, is equal to 2 (thus its nucleus has two protons); it has two isotopes, He^3 and He^4.

The 325 different nuclei are combined into 104 elements, of which ninety-two are called natural (their atomic numbers vary from 1 for hydrogen to 92 for uranium) and twelve are called artificial because they have been obtained only in the laboratory.

In the terrestrial crust the most abundant elements are oxygen (O^{16}) and silicon (Si^{28}), but there are a great many others in significant proportions. In the known universe as a whole, however, it is observed that there is 75 percent of H^1 and 24 percent of He^4, all other elements representing only 1 percent of the total.

Atoms may be combined to form certain molecules that are characteristic of chemical compounds. Thus, two atoms of oxygen form the oxygen molecule, which is represented by O_2; it is in this form that gaseous oxygen is found in the atmosphere. The water molecule is composed of two atoms of hydrogen and one atom of oxygen (the chemical formula H_2O). The stability of molecules is much less than that of nuclei, for the atoms of which they are composed are bound by electrons situated on the periphery of the nuclei. The dissociation of one molecule into other molecules or the association of several molecules to form other molecules constitute what are referred to as "chemical reactions."

d2. RADIOACTIVITY. The binding energy of the different particles in the nucleus of an atom is very great. However, certain nuclei disintegrate spontaneously and this disintegration releases energy, either in the form of particles ejected at very high velocities, or in the form of photons without electrical charge or mass.

Alpha Particles. Certain heavy unstable nuclei will suddenly eject a particle composed of two protons and two neutrons, which is called an alpha (α) particle. Let us note that such a particle is none other than a nucleus of helium four (He^4). The speed of ejection is such that the kinetic energy of the α particle is considerable. As for the nucleus remaining after ejection of the α particle, it no longer contains the same number of protons and neutrons as previously; thus there has been a transmutation from one element to another. As an example, let us select polonium 210 ($Z = 84$, $A = 210$), which can eject an α particle; it is then transformed into lead 206 ($Z = 82$, $A = 206$). This is expressed

$$ {}_{84}\mathrm{Po}^{210} \rightarrow {}_{82}\mathrm{Pb}^{206} + {}_{2}\mathrm{He}^{4} + 5.3 \text{ Mev} $$

Mev stands for "mega electron-volt" and it represents the energy released; it is equal to one million electron-volts, an electron volt being the energy acquired by one electron in an electrical field, accelerated through a potential difference of one volt.

Beta Radiation. It may happen that a neutron of a nucleus will disintegrate while ejecting an electron; the neutron is transformed into a proton (the sum of the electrical charge ejected and the remaining charge is zero). The emission of these electrons constitutes beta (β) radiation. For example:

$${}_{82}Pb^{210} \rightarrow {}_{83}Bi^{210} + 1 \text{ electron}$$

Lead 210 has been transmuted into bismuth 210.

Gamma Radiation. It sometimes happens that in the process of an α or β disintegration the nucleus may retain a portion of the energy released by the disintegration at the expense of the particle ejected. This energy of the "excited" nucleus is then released in the form of photons (luminous particles characterized by electromagnetic radiation at high frequencies) endowed with great energy. This is known as gamma (γ) radiation. This phenomenon also occurs with "excited" atoms or when particles travel through matter; it is then given the name of X radiation, or X rays. This natural phenomenon can be reproduced in many ways in the laboratory, which has made it possible to understand the structure of nuclei.

Generally speaking, the α particles ejected from a nucleus encounter other atoms and, because of their positive charge, they "rob" the electrons from these atoms. α particles have, therefore, an ionizing effect. However, it may happen that the α particles may strike against a nucleus and steal protons from it (hydrogen nuclei), thus releasing a large quantity of energy; this is an example of those reactions referred to as nuclear reactions.

Finally, attention should be called to another group of reactions, called fusion reactions, in which nuclei are recombined to form heavier nuclei with a release of energy; this phenomenon will play an important role in the theory regarding the energy of the stars. Thus two isotopes of hydrogen may produce a nucleus of helium 4 while releasing a neutron (as in the hydrogen bomb).

d3. THE COMPOSITION OF THE ATOM; SPECTRA. At the beginning of this century the atom was regarded as a miniature solar system with the electrons rotating around the nucleus like planets around the sun. However, this interpretation has had to be revised; electrons are now regarded sometimes as particles, and sometimes as waves. The position of an electron on its orbit cannot be determined with precision, like the position of a planet, but obeys a certain probability law.

The electrons around the nucleus can only be found in certain shells called, starting from the nucleus, *K*, *L*, *M*, etc., through *Q* (there are a maximum of seven shells); each shell is itself divided into subshells. An electron may jump from one shell to another; if it is separated from the nucleus, it may provide energy. Each shell possesses a certain clearly defined level of energy and the different levels of energy can only acquire certain values, without the possibility of continuously changing the level of energy. This theory is the subject of quantum physics.

If an atom passes from an energy level E_1 to a lower energy level E_0 (one electron having jumped to a shell closer to the nucleus), there occurs a release of energy in the form of an electromagnetic radiation of wavelength λ (or of frequency $\nu = 1 / \lambda$) and we have

$$E_1 - E_0 = h\nu$$

h being a constant known as Planck's constant. An atom may also "absorb a frequency." This means that if the atom passes to a higher energy level (here E_0 is greater than E_1), this is caused by the absorption by the atom of a quantity of energy equal to $E_0 - E_1$, to which there corresponds a frequency

$$\nu = \frac{E_0 - E_1}{n}$$

The spectrograph is an optical device with which one can separate the radiation of various wavelengths proceeding from a light source. It consists of a small rectilinear slit through which light passes; a collimating lens, which makes the light rays parallel; and a prism, which deflects light rays differently according to their wavelength. One receives on a photographic plate different rectilinear images of the slit, which are called lines; the relative positions of the lines correspond to the wavelengths of the light that forms them. A group of lines as a whole constitutes the spectrum of the source.

A rarefied gas that has been excited (for example, neon in a light tube) has a spectrum composed of fine brilliant lines; these are emission lines. If one interposes in front of this excited gas some gas which is not excited, one observes in place of the brilliant lines dark absorption lines. Each line is characteristic of the element and enables one to identify it and to determine the state of excitation of the atoms. Molecules produce spectra composed of lines gathered into rather wide bands.

In hydrogen—the simplest case, since the hydrogen atom has only one electron—the lines of the spectrum lend themselves to simple expression. If λ is the wavelength of one line, we have

$$\frac{1}{\lambda} = R\left(\frac{1}{m^2} - \frac{1}{n^2}\right)$$

m and n are whole numbers ($n > m$) and R is a constant known as Rydberg's constant. If we assume that $m = 2$ and if we cause n to vary, we obtain the Balmer series, whose lines are situated in the visible area of wavelengths. As n becomes larger, the lines move closer together and eventually merge in what is called the continuous spectrum.

When $m = 1$, we have the Lyman series, whose lines are in the ultraviolet area; when $m = 3$, we have the Paschen series, etc.

d4. THE STARS: STRUCTURE AND EVOLUTION. Spectral study of the light of the stars has made it possible to classify them in certain categories which correspond to the temperature of the star. The hottest ones correspond to class O, characterized by the presence of lines of ionized helium. After this we have class B (no more ionized helium, appearance of lines of hydrogen), class A (no more helium, lines of hydrogen and calcium), class F (appearance of lines of ionized metals), class G (appearance of lines of neutral metals), class K (neutral metals and molecular bands), and class M (bands of titanium oxide). The stars become increasingly cool as one passes from class O to class M. Similarly, the color of the star changes, passing from blue (O stars) to red (M stars), by way of yellow (G stars).

By measuring the distances of the stars one can, when one knows their apparent magnitudes, obtain their absolute magnitudes, which express the star's intrinsic luminosity, that is, the quantity of energy it radiates.

The Hertzsprung-Russell diagram is constructed as follows: one uses as the abscissa the spectral class (from O to M) and as the ordinate the absolute magnitude. (Let us bear in mind that the lower its magnitude, the more brilliant the star.) A star is represented by a point on this graph (Fig. 64).

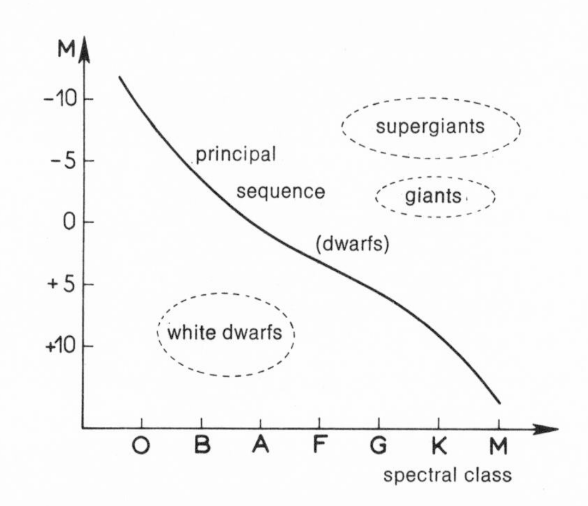

FIGURE 64

It was observed that most of the points are grouped around one continuous line. The corresponding stars are called dwarfs or stars of the main sequence. With stars of this kind, the bluer they are, the more brilliant they are. The sun is a star of the main sequence of class G (yellow stars) with an absolute magnitude of around +5. Outside the main sequence, there exist other important groups of stars: the white dwarfs, which are dim but of spectral type B, A, or F, corresponding to high temperatures; the giants, which are red and consequently cold, but very brilliant; and the supergiants, which are even more brilliant.

We now know that the energy of the stars is due to nuclear reactions; in particular hydrogen is transformed into helium, either by the process known as the Bethe cycle, in which the nucleus of carbon 12 (C^{12}) plays an intermediary role, or by the chain process called the proton-proton chain, in which intermediary reactions cause the appearance of nuclei of deuterium and helium 3.

The idea that now prevails about the evolution of the stars is as follows: there is still little certainty about the birth of a star, which is generally attributed to the contraction of a cloud of matter. Afterward, luminosity decreases while temperature remains constant; then the representative point of the star follows the main sequence, as luminosity increases; finally, when its hydrogen has been exhausted, the star evolves rapidly as it reaches the realm of the giants or supergiants (Fig. 65).

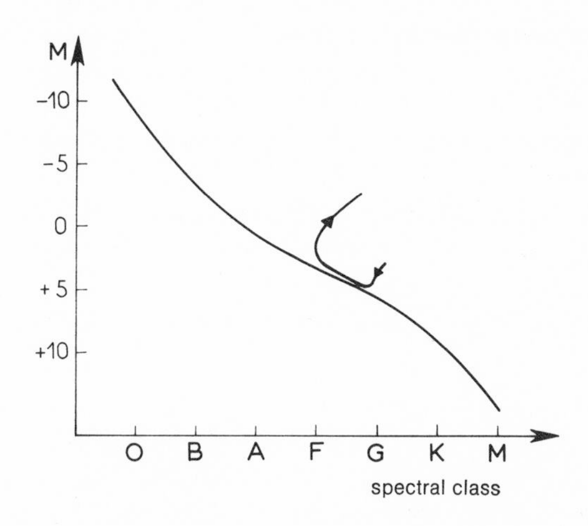

FIGURE 65

As for the white dwarfs, these are extremely dim but hot stars; their density is extremely high. It is believed that they represent a late state of stars which have exhausted all the resources of nuclear energy, and in which the nuclei of the atoms are densely packed together.

II. DESCRIPTION OF THE UNIVERSE

a. VARIABLE STARS AND CRITERIA OF DISTANCE

a1. GENERAL REMARKS. The luminosity of certain stars is not constant. We know tens of thousands of variable stars which have been discovered either by visual observation or by the comparison of photographs of the same region of the sky taken at different periods. Some of these stars have variations which are not periodical; this is the case with the novae and supernovae, whose luminosity suddenly increases and then slowly diminishes. The most celebrated supernova is the one that appeared in 1054, and which was visible in broad daylight; it left behind a gaseous nebula, the Crab Nebula. Other stars, termed eruptive, present sudden increases in luminosity less substantial than those of the novae and supernovae; examples of these are stars of the T Tauri or UV Ceti type. With stars of the so-called R Coronae Borealis type, luminosity will suddenly diminish. Finally, certain stars have variations in luminosity which are more or less periodical, but rather irregular. Stars of the Mira Ceti type have a long cycle (between 100 and 1,000 days) but one which is not absolutely constant. A star such as RV Tauri has a cycle of 78 days but highly irregular variations in amplitude.

a2. PERIODIC VARIABLES. Regular periodic variables with short cycles will be most interesting to us, since, as we shall see, they play an important role in the judging of distances.

The typical star of this kind is the star Delta Cephei, which gives its name to all the so-called Cepheid stars. The magnitude of this star varies from 4.1 to 5.2, then back to 4.1 in a highly regular manner, within a period of 5.37 days; the decrease in luminosity is a little slower than the increase. Delta Cephei is a pulsating star whose volume increases and decreases; this is accompanied by a variation in temperature which causes a variation in luminosity. Cepheids are found in our galaxy, in the vicinity of its plane of symmetry, called the galactic plane; but they are also found in the Magellanic Clouds, which are small irregular galaxies gravitating around our own, as well as in other extragalactic nebulae.

Stars of the RR Lyrae type are also very regular periodic variables, but their cycles are very short, less than one day. The light curve is not identical to that of the Cepheids, as we see in Fig. 66, since their magnitude remains almost constant in the vicinity of their minimum luminosity for a certain period of time.

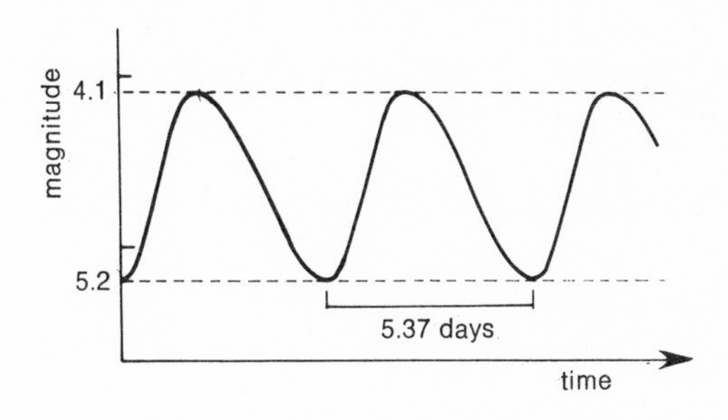

Cepheids

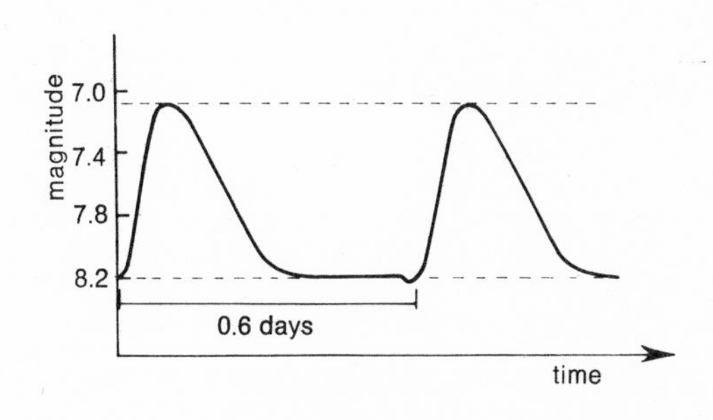

R R Lyrae

FIGURE 66

Stars of the RR Lyrae type are found in our galaxy, but they are distributed more generally and are not concentrated near the galactic plane. They are also found in the globular clusters, which are spherical groups of stars revolving just within the confines of our galaxy.

All these stars (Cepheids and stars of the RR Lyrae type) are of spectral types A, F, G, or K. Generally speaking, the longer its cycle, the colder the star; the Cepheids, for example, are of type F when their cycle is 1.5 days, and of type K when their cycle is 33 days.

a3. THE RELATIONSHIP OF CYCLE TO LUMINOSITY. This important relationship was discovered by Henrietta Leavitt in 1912. It has since been improved on; specifically, we have learned to make better use of the fact that a star is a Cepheid or a star of the RR Lyrae type. Generally speaking, the relationship indicates that the cycle is longer when the average absolute magnitude of the star is greater, which means that a star which is intrinsically very brilliant will have a longer cycle than a dimmer star. As far as stars of the RR Lyrae type are concerned, their absolute magnitude is, on the average, practically constant.

Figure 67 illustrates this relationship. On it we see that absolute magnitude is practically a linear function of the logarithm of the period (expressed in days). The absolute magnitude of stars of the RR Lyrae type remains close to zero. The straight

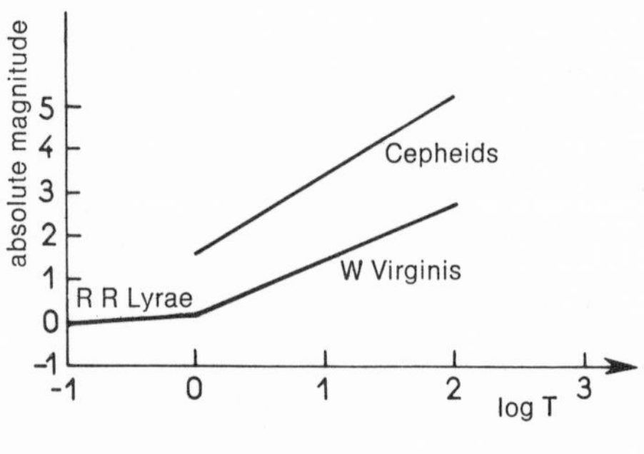

FIGURE 67

line below the one representing the Cepheids corresponds to stars of the W Virginis type, also called abnormal Cepheids.

We see the vast importance of the relationship between period and luminosity: when one has recognized by its light curve that a certain star is a periodic variable of a certain type with a certain period, one can deduce its absolute magnitude M and consequently its distance r by means of the formula

$$m = M - 5 + 5 \log r$$

m being apparent magnitude. The quantity $m - M$ is called the distance modulus.

b. STRUCTURE OF THE GALAXY

b1. GENERAL APPEARANCE. The sun is a member of a vast complex comprising about two hundred billion stars called the galaxy. This complex has the shape of a flattened disc, containing approximately 70 percent of the stars in the galaxy and the interstellar gases, surrounded by a halo containing the rest of the stars. The total mass of the galaxy is approximately 2.5×10^{11} solar masses.

The radius of the disc is approximately 20,000 parsecs (65,000 light-years); the sun is located near the plane of symmetry of the disc, at about 10,000 parsecs from the galactic center. The Milky Way is the image produced on the celestial vault by all the stars concentrated in the disc (Fig. 68).

The galaxy contains groups of stars of irregular shape called open clusters, which are generally found concentrated near the plane of symmetry. Other clusters, very rich in stars, gravitate at the outer edges of the halo; these are called globular clusters because of their spherical shape and because of the high concentration of stars at their centers. The best known of the open clusters are the Pleiades and the Hyades; Hercules, which appears to the naked eye as a star of the fourth magnitude, is a globular cluster which can be resolved into separate stars with a modest instrument.

The space between the stars of the galaxy is not empty; it

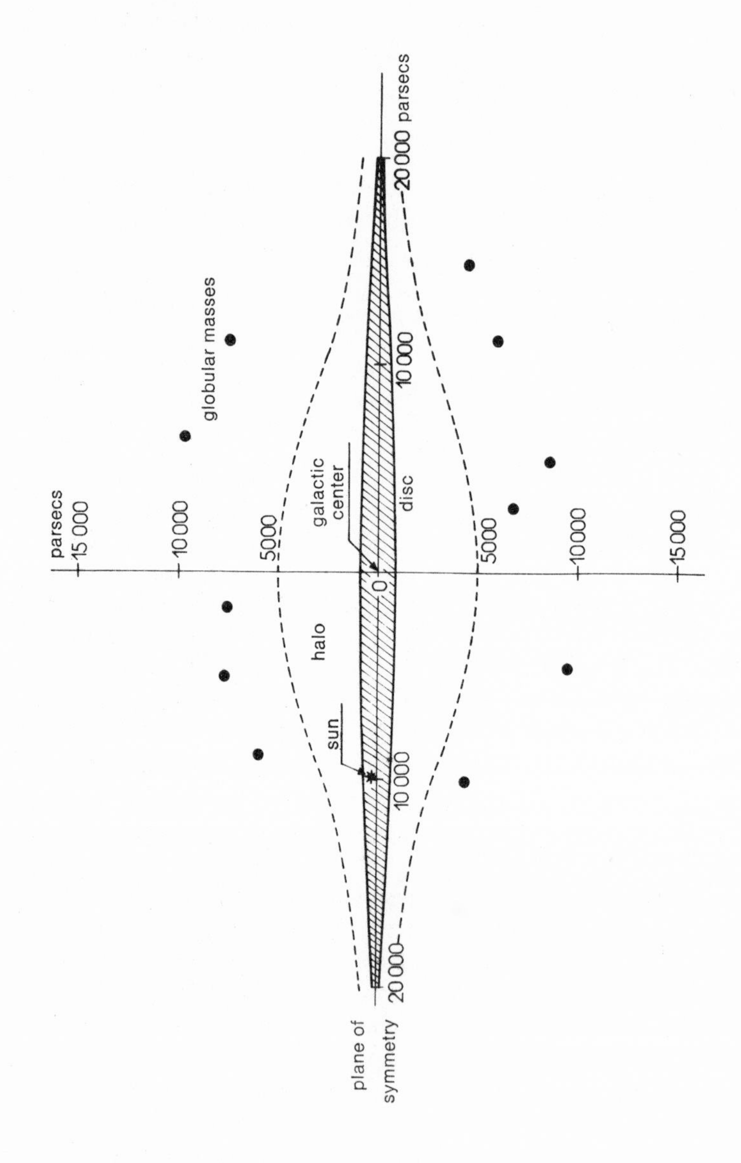
parsecs
15 000
10 000
5000
0
5000
10000
15 000
globular masses
galactic center
disc
halo
sun
10 000
10 000
20 000 parsecs
20 000
plane of symmetry

FIGURE 68

contains interstellar matter composed of gas and dust. In addition to the arms of hydrogen of the galaxy, discussed later, there are the dark nebulae which hide the stars; they are numerous in the vicinity of the galactic plane. Other nebulae, however, are brilliant; they are either diffuse, like the nebulae of Cygnus, or relatively dense and of spheroid form surrounding a hot star (planetary nebulae).

b2. DIFFERENTIAL ROTATION. The galaxy rotates on an axis that passes through its center and is perpendicular to the galactic plane, but this rotation is not uniform at every point. The region of the center rotates like a solid body, but the more remote regions have velocities lower than the ones they would have if they were rotating in the same manner. For example, in the vicinity of the sun, if ω is the angular speed of rotation, r the distance from the center, and v velocity, we have

$$\omega = \frac{v}{r} = A - B; \quad \frac{dv}{dr} = A + B$$

A and B are two constants called Oort's constants. A equals 15 kilometers per second per kiloparsec (1 kiloparsec = 1,000 parsecs) and B equals -10 kilometers per second per kiloparsec. Thus we have

$$\omega = 25 \text{ km/sec/kpc}; \quad \frac{dv}{dr} = 5 \text{ km/sec/kpc}$$

Since the sun is approximately 10,000 parsecs from the center of the galaxy, we see that its speed is 250 kilometers per second and that it would vary by 5 kilometers per second for a variation of distance from the center of the galaxy of 1,000 parsecs. This phenomenon is known as the differential rotation of the galaxy.

However, the phenomenon described above is an average phenomenon; since each star is impelled by an individual motion, it is only on the average that the complex of stars near the sun participates in differential rotation.

b3. RADIOASTRONOMY; HYDROGEN CLOUDS. Our understanding of

the galaxy and indeed of the universe has been greatly improved since a rather new technique, radioastronomy, was developed about twenty years ago.

The matter that constitutes warm bodies emits an electromagnetic radiation caused by the agitation of free electrons within these bodies. The light received from the stars by means of telescopes or by the naked eye is an electromagnetic radiation of very short wavelength, on the order of a few tenths of a micron (one micron is equal to 10^{-6} meters). The stars are also emitting over much longer wavelengths, varying from a few centimeters to several meters. However, long waves are absorbed by the ionosphere, so that we shall be limited to wavelengths shorter than twenty meters. Similarly, absorption by the molecules of the atmosphere limits observation of radio waves of too short wavelengths.

The basic instrument of radioastronomy is the radiotelescope, consisting of a spherical or parabolic mirror made with a wire netting whose meshes must be smaller than one-fourth of the wavelength being studied. At the focus of the mirror there is a bipolar antenna which receives the signals. Its resolving power depends on the diameter of the mirror, and it is very difficult to construct mirrors of this size which can be swiveled. At the Observatory at Nançay, the mirror is fixed, though it is a plane mirror that has a sufficient degree of freedom in the mechanical sense to permit scanning in zenithal distance.

Radiotelescopes are also constructed in the form of interferometers consisting of two radiotelescopes of reduced size bound to the same receiver. The wave arriving from a radio source produces interference fringes caused by the difference in path, the width of a band being proportional to λ/D (λ is the wavelength, D the distance from the two radiotelescopes to the interferometer). Another technique is to use networks formed of identical equidistant antennae lined up and bound to a receptor. The one at Nançay is 1,550 meters in length and works for wavelengths of approximately 1.8 meters.

Study of the galaxy by means of radioastronomy has proved to be extremely fruitful. This technique has made it possible to identify radio sources with the remnants of supernovae, the emission being caused by electrons of high energy in a magnetic field.

An extremely interesting radio emission is the radiation of hydrogen over a wavelength of 21 centimeters. This radiation is caused by the transition from one atomic state to another, these states being characterized by the relative directions of the spins of the electron and the proton making up the hydrogen atom. (The spin of the electron or proton consists of its rotation about an axis.)

The hydrogen line of 21 centimeters is not encountered on the earth; it has only been produced artificially in the laboratory (such lines are referred to as forbidden lines). In interstellar space, the 21-centimeter line is emitted by neutral interstellar hydrogen and provides very important information about the structure of the galaxy. Vast clouds of hydrogen have been discovered and they seem to be arranged in arms coiled in the shape of spirals, as indeed, are the stars themselves; this spiral structure, which we shall encounter later in studying other galaxies, has recently been established with certainty.

In the study of the extragalactic universe, radioastronomy plays a role of major importance; a great many extragalactic radio sources have been detected. Occultations of radio sources by the moon and long-baseline interferometry make it possible to localize these sources with great precision.

c. GALAXIES

We are concerned here with objects sometimes called extragalactic nebulae or external galaxies. They are agglomerations of stars—several hundred billion—and gas, usually very similar to our own galaxy and distributed throughout the universe in a homogeneous and isotropic manner.

There exist galaxies about 20,000 parsecs in diameter, containing two or three hundred billion stars, but also much smaller galaxies, 4,000 parsecs in diameter, containing about ten billion stars.

c1. SHAPES OF GALAXIES; HUBBLE'S CLASSIFICATION. Galaxies appear in a telescope as small pale spots; some are visible to the naked eye, like the Magellanic Clouds or the Andromeda nebula M 31 (i.e., number 31 in Messier's catalogue). For a long time people wondered whether or not these objects belonged to our galaxy, until more powerful telescopes made it possible to resolve them into stars and to calculate their distances.

The different types of galaxies are distinguished by their shapes. Some have a completely irregular shape, like the Magellanic Clouds; others have the shape of a somewhat flattened ellipsoid (elliptical nebulae); still others have a lenticular shape in which one can distinguish a spiral structure that is more or less clear (normal spiral galaxies). In certain spiral galaxies, the whorls seem to begin at either end of a bar that passes through the center (barred spirals). Figure 69 illustrates the possible shapes of galaxies.

Hubble invented the following classification. Elliptical galaxies are designated by E, with a subscript varying from 0 to 7, which increases as the galaxy becomes flatter. Galaxies of lenticular shape will be designated by SBO or SO according to whether

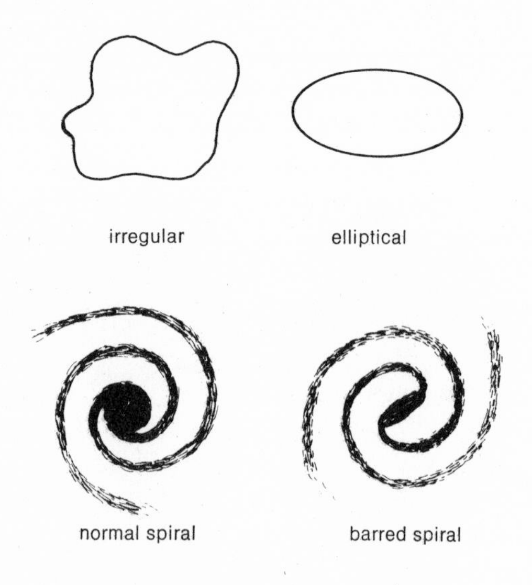

FIGURE 69

or not they possess the beginning of a bar as they approach the spiral form (subscript varying from 1 to 3). The spirals are designated by S and the barred spirals by SB, and each series has distinct types, a, b, and c, in which the arms are further and further open. Finally, irregular galaxies are designated as "Irr." One can place any galaxy in the diagram (Fig. 70).

The spirals, barred and otherwise, represent approximately 61 percent of those galaxies whose magnitude is lower than 12.7, according to Gerard de Vaucouleurs, elliptical galaxies 13 percent, irregular galaxies 2.5 percent, and SO's 21.5 percent.

According to the latest estimates, the masses of the galaxies are on the order of 10^{10} times the mass of the sun (that is, ten billion solar masses). The largest galactic mass known is that of the galaxy M 87 (Messier 87), which is said to be 2.7×10^{12} times that of the sun (approximately three trillion times the mass of the sun). The mass of the Andromeda nebula is estimated to be 4.3×10^{11} solar masses.

We are still far from knowing all the galaxies that our instruments should enable us to reach. In fact, although tens of thousands of galaxies have been identified, it is estimated that there are approximately 6×10^{8} with magnitudes brighter than 23. It is believed that there are about 10^{-23} galaxies per cubic parsec in the universe and that the density of their matter is 10^{-30} grams per cubic centimeter.

c2. DISTANCES OF THE GALAXIES. We have seen that the discovery of Cepheids in the Magellanic Clouds enabled Leavitt, in 1912, to discover the relationship between period and luminosity. In 1923, Cepheids were discovered in the Andromeda galaxy (M 31), thus enabling us to calculate the enormous distance that separates us from it. This first measurement decided definitively in favor of the theory that regarded extragalactic nebulae as objects analogous to our galaxy and distributed in space at great distances from one another. The distance of the Andromeda nebula is 2,000,000 light-years. If one wishes to estimate the distances of galaxies beyond 10,000,000 light-years, for example, the Cepheids are too dim to be observed; and one must direct one's attention to blue supergiant stars whose absolute magnitudes one will assume to be the same as those of identical stars in the Andromeda galaxy. To calculate the distances of galaxies so remote that one cannot distinguish their stars, one must try to discover the absolute magnitudes of the galaxies themselves, but the problem is complicated by the spectral shift, of which we shall speak further on. Hubble's Law, studied in the following paragraph, is also used to determine the distances of remote galaxies.

c3. THE DOPPLER EFFECT; THE EXPANSION OF THE UNIVERSE. Let us consider a point O and a light source A which is moving away from or toward O at velocity v, the vector velocity of A with respect to O being expressed by OA. (In this case we say that the velocity of A in relation to O is radial.) If λ is the wavelength of the light emitted by A, the wavelength received by O will not be equal to λ but to $\lambda \pm \Delta\lambda$, with

$$\frac{\Delta\lambda}{\lambda} = \frac{v}{c}$$

c being the speed of light. For example, if v is one-thousandth of the speed of light (300 kilometers per second) and if the light emitted by A is a line of sodium whose wavelength is 5,890 Å [1 Å (angstrom) = 10^{-8} cm], the light received at O will have a wavelength of (5,890 + 5.89) Å, if A is moving away from O; and (5,890 − 5.89) Å, if A is moving toward O.

This effect is known as the Doppler effect. The light received is shifted toward red (long wavelength) if the source is moving away from the observer, and toward violet (short wavelength) if the source is moving toward the observer.

This effect is in fact common to every vibratory phenomenon. If A is emitting a musical sound of a given frequency, an analogous shift occurs; in this case one must replace c in the above formula with the speed of sound in air.

Getting back to light sources, let us assume that the velocity v of A is now a substantial fraction of the speed of light c. The formula given no longer applies because of relativistic considerations, and it has been shown that it must be replaced by the formula

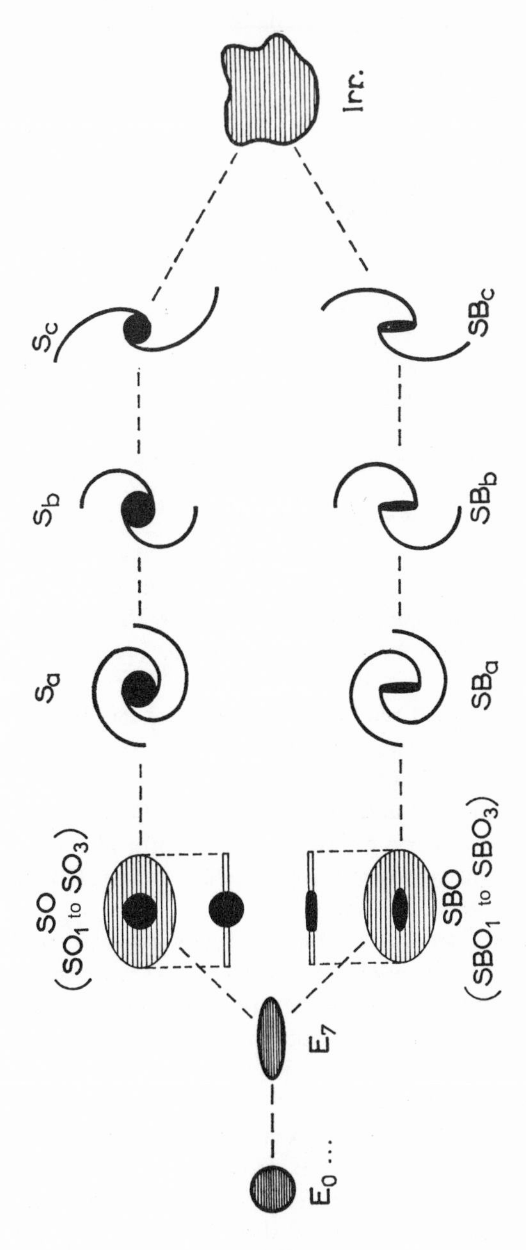

FIGURE 70

$$\frac{\Delta\lambda}{\lambda} = \frac{1 + v/c}{\sqrt{1 - v^2/c^2}} - 1$$

If v/c is very small, $\sqrt{1 - v^2/c^2}$ is close to one and we again find the formula given above; by disregarding v^3/c^3, one can write

$$\frac{\Delta\lambda}{\lambda} \cong \frac{v}{c} + \frac{1}{2}\frac{v^2}{c^2}$$

If $v/c = 1/10$ and $\lambda = 5{,}890$ Å,

$$\Delta\lambda = [5{,}890 \pm (589 + 29.45)] \text{ Å}$$

the sign being + if *A* is moving away from *O*, and − if it is moving toward *O*.

Now let us imagine certain atoms or molecules that are emitting a spectrum of clearly defined lines. If, on a spectrograph, a spectrum emitted by a star is absolutely identical to a known spectrum, except that all the lines have shifted by an amount proportional toward red or toward violet, it can be deduced that the star is moving away from or toward the observer and one will be able to calculate its radial velocity v.

This can also be done, of course, with the spectra of galaxies. It is in this way that we find that the spectra of the Andromeda nebula have shifted toward violet, which proves that this galaxy is moving toward our own.

However, an extraordinary discovery was made rather quickly, once the exploration of the universe and the study of the spectra of numerous galaxies had reached a certain point. For all remote galaxies, without exception, the shift in the spectra occurs toward red; moreover, this shift is greater if the galaxy being studied is more remote.

If the shift is always interpreted as a Doppler effect, it would appear that galaxies move away from one another with velocities that increase with distance. The fact that the spectrum of the Andromeda galaxy exhibits a shift toward violet rather than toward red is not in contradiction with what has just been said. In effect, this galaxy is relatively very close to ours, and its own motion takes precedence over the motion caused by the expansion of the universe.

The Hubble-Humason law expresses the relationship between the speed of recession of a galaxy and its distance. This relationship is linear and we have

$$v = Hr$$

r being the distance of the galaxy in question, v the radial velocity of its recession, and H Hubble's constant, equal to 75 kilometers per second per megaparsec (one megaparsec equals one million parsecs). The number $1/H$ expresses the interval of time it has taken a galaxy to travel distance r at velocity v, and this is true no matter what galaxy is being considered. In other words, if the expansion is real and if it has always occurred according to the Hubble-Humason law, $1/H$ represents the time that has elapsed since the beginning of the expansion; $1/H$ is equal to 13.5 billion years. This number is sometimes referred to, rather improperly, as the age of the universe.

We shall return to the interpretation of the spectral shift in section III.

d. OBJECTS AND PHENOMENA RECENTLY DISCOVERED IN THE UNIVERSE

d1. GALAXY CLUSTERS. Very often, galaxies are found grouped into clusters which sometimes include several tens of thousands of members. These clusters generally have a spheroid form. With respect to the near extragalactic universe, it has recently been recognized that about a hundred groups each comprising ten galaxies form a system centered around the important cluster of Virgo, which is located ten megaparsecs from our galaxy. It is thought that our galaxy is located at the outer edge of a vast cluster of galaxies centered around the cluster of Virgo which is called the local supercluster.

d2. QUASARS. These objects were discovered very recently in the form of pointlike radio sources which have been associated with a luminous object similar to a star. The word quasar is a contraction of the term quasi-stellar. These objects exhibit a very pronounced shift in their spectrum toward red ($\Delta\lambda/\lambda$ exceeds 2), and we now have every reason to believe that this shift is caused by the expansion of the universe and that, consequently, quasars are very remote. But in this case, given the large amount of energy that we receive from them in spite of their distance, quasars must at every moment release a fantastic amount of energy which no known physical phenomenon enables us to explain as yet.

Spectroscopic observation of quasars reveals the presence of ionized magnesium, triply ionized carbon, and the Lyman α line of hydrogen. All these lines have wavelengths that are usually much too short to be visible, but the very substantial shift toward red of the spectra of quasars brings these rays into an accessible area of the spectrum.

d3. PULSARS. These are radio sources whose frequency of emission varies periodically with remarkable regularity; the cycle of pulsars falls between 0.03 second and about four seconds. Pulsars are galactic objects often associated with gaseous nebulae, the remnants of supernovae. This is true, for example, of the pulsar discovered in the Crab Nebula. Pulsars are now believed to be stars composed of neutrons, whose axis of rotation forms a certain angle with the axis of the magnetic moment.

d4. COSMOLOGICAL RADIATION. Scientists have discovered a very weak electromagnetic radiation which appears in a continuous and isotropic manner in the universe. This radiation is characteristic of a "black body," that is, it resembles the radiation of a warm body that found itself in this case at a temperature of −270°C, whereas the lowest temperature one can reach (absolute zero) is −273°C. The only explanation now being considered for this radiation is that it is the result of a gradual, but not yet completed, cooling that may have followed an initial explosion at the beginning of the expansion of the universe.

III. MODERN COSMOLOGY

a. COSMOLOGICAL MODELS

a1. GENERAL IDEA. The purpose of cosmology is to study the universe as a whole, especially the distribution of matter and of electromagnetic or gravitational fields created by this matter.

Classical mechanics referred, of course, to the notion of space and time in trying to establish physical laws, but it endowed these concepts with a character altogether distinct from that of the objects of physics. Space and time could be conceived independently of any matter or of any field. Moreover, Euclidean space was conceived to be three-dimensional and infinite; the time that flows uniformly from minus infinity to plus infinity was in no way related to it.

Special relativity preserves the same state of mind, except that the frame of reference of physics is no longer merely space—absolute time now serving as a parameter to locate phenomena—but a Euclidean space-time of four dimensions. General relativity contributes some totally new ideas to physics, in that the space-time in the vicinity of matter exists only by virtue of this matter. Let us be more specific: we have seen that, given a certain distribution of matter, general relativity seeks to determine the metrical tensor of components twice covariant $g_{\alpha\beta}$. In some sense, there is no space-time in the vicinity of matter unless there is matter (or a field).

We have stressed the fact that the theory of general relativity was a local theory, because the identification between the forces of gravitation and the forces of inertia is valid only at a finite distance. However, the equations of general relativity, to which we shall return, relate for a given point the distribution of matter to the coefficients of the metrical tensor, and it is conceivable that these equations could be integrated for the entire universe, given certain conditions, within whatever limits it will be appropriate to set.

One could attempt to solve the problem of cosmology within

the frame of reference of classical mechanics, but it is obvious that enormous difficulties present themselves from the outset. In the first place, the interpretation of the red shift of the spectra of remote galaxies as a motion of expansion indicates that velocities that are not negligible in relation to the speed of light exist in the universe; and we know that in this case classical mechanics is inadequate to accurately describe reality. Moreover, the intense gravitational fields that prevail in galaxies cannot really be interpreted except in terms of general relativity. (Classical mechanics is inadequate even to represent the motion of Mercury.) In this connection it should be noted that the red shift of the spectra of remote galaxies cannot be attributed to the fact that the sources of emission are located in intense gravitational fields; indeed, if this were the case, the gravitational potential would become increasingly high in proportion as one moved away from our galaxy, which would be quite extraordinary and would seem to give our galaxy a unique and privileged situation, which is contrary to modern scientific thought.

It is agreed that, apart from local anomalies, the universe is such that matter is distributed uniformly and in an isotropic manner for any observer. It should be noted that this presupposes the existence of a universal (or absolute) time in relation to which, at a given moment, the metric of space is the same at every point. This metric may in fact evolve in the course of this time; it is in this way that we shall interpret the expansion of the universe.

Consequently, space (i.e., three-dimensional space) has everywhere an identical curvature which may be positive, negative, or zero. This curvature may, moreover, be a function of absolute time.

a2. EINSTEIN'S EQUATIONS. The equations of Einstein's general relativity are presented in the form of equivalences between the components of two tensors. One of these tensors is related to the geometry of space-time; it depends only on the metrical tensor. Einstein was led to take the tensor of twice covariant components $S_{\alpha\beta}$ which are such that:

$$S_{\alpha\beta} = R_{\alpha\beta} - \tfrac{1}{2} g_{\alpha\beta} (R - 2\lambda); \quad (\alpha, \beta = 1, 2, 3, 4)$$

The $g_{\alpha\beta}$'s are the twice covariant components of the metrical tensor; $R_{\alpha\beta}$'s the components of a tensor called Ricci's tensor, which is wholly determined when one knows the $g_{\alpha\beta}$'s; R is a scalar quantity resulting from Ricci's tensor. As for λ, it is a constant, called the cosmological constant, which Einstein did not introduce into his equations until several years after he published them. In fact, as we shall see further on, it was impossible for him to find as the solution to his equations a static universe if λ was zero. However, contemporary cosmologists tend to assume that $\lambda = 0$, which is, in fact, perfectly compatible with models of an expanding universe.

Assuming this, we shall write

$$S_{\alpha\beta} = -8\pi G T_{\alpha\beta}$$

$-8\pi G$ being a constant of proportionality (G is the constant of universal gravitation), and $T_{\alpha\beta}$ twice the covariant components of a tensor that expresses the state of distribution of matter or energy at the point of space-time in question, whereas $S_{\alpha\beta}$ is a purely geometrical tensor. In the absence of any matter, the $T_{\alpha\beta}$'s would be zero. Einstein's equations will be written

$$R_{\alpha\beta} - \tfrac{1}{2} g_{\alpha\beta}(R - 2\lambda) = 0 \qquad (1)$$

or

$$R_{\alpha\beta} - \tfrac{1}{2} g_{\alpha\beta}(R - 2\lambda) = -8\pi G T_{\alpha\beta} \qquad (2)$$

Equation (1) will be utilized where there is neither matter nor field. These equations should be compared with Laplace's and Poisson's equations in classical mechanics.

If $U\ (x, y, z)$ is the gravitational potential at a point whose coordinates are x, y, z, we have

$$\frac{\partial^2 U}{\partial x^2} + \frac{\partial^2 U}{\partial y^2} + \frac{\partial^2 U}{\partial z^2} = 0$$

or

$$\frac{\partial^2 U}{\partial x^2} + \frac{\partial^2 U}{\partial y^2} + \frac{\partial^2 U}{\partial z^2} = 4\pi G_\rho$$

$\frac{\partial^2 U}{\partial x^2}$, etc. being the partial derivatives of the second order of U in relation to x, etc. The first equation—Laplace's equation—is valid outside any mass; the second—Poisson's equation—is valid inside masses. G is the constant of universal gravitation and ρ is density. It would be difficult to show here how Einstein arrived at equations (1) and (2); we can only say that he tried to find in general relativity equations that had properties analogous to the equations of Laplace and Poisson.

The tensor $T_{\alpha\beta}$ is called the energy-momentum tensor; it is a function of the density of matter and of pressure p as well as of the components of the vector velocity of a particle. The weakness of the theory from the point of view of cosmology is that tensor $T_{\alpha\beta}$, as we know how to construct it, does not include all the physical properties of matter; in particular, it does not account for the quantum nature of energy. The universe is reduced to a uniform distribution of matter having the properties of a perfect gas.

a3. THE ROBERTSON-WALKER $\overline{ds}^2$. By utilizing coordinates analogous to the spherical coordinates r, θ, ϕ for space, it can be demonstrated that the most general ds^2, corresponding to a homogeneous and isotropic universe, may be expressed in the form

$$\overline{ds}^2 = c^2\overline{dt}^2 - \frac{R^2(t)}{\left(1 + \frac{Kr^2}{4}\right)^2}(\overline{dr}^2 + r^2\overline{d\theta}^2 + r^2 \sin^2 \theta \overline{d\phi}^2)$$

K is a constant equal to $+1$, 0, or -1, which is such that K/R^2 represents the curvature of space, constant at every point for a given moment, since R is solely a function of t. If K is zero, the $\overline{ds}^2$ for t constant ($dt = 0$) is a three-dimensional Euclidean $\overline{ds}^2$, space is Euclidean, its curvature is zero. If $K = 1$, the curvature is positive, the space (at t constant) is a closed space with a constant curvature, analogous to a sphere; it is a hypersphere having three dimensions which one can imagine immersed in a space of four dimensions (which has nothing to do with space-time). Finally, if $K = -1$, the curvature is constant and negative. In all cases, the function $R(t)$ is what is known as the radius of the universe; this function $R(t)$ should not be confused with the quantity R which appears in the first member of Einstein's equations and which is called a "Riemannian scalar curvature."

Now let us introduce the coefficients $g_{\alpha\beta}$ of the Robertson-Walker $\overline{ds}^2$ into Einstein's equations. What we are doing, since we are unable to solve these equations in the most general case, is seeing what equations verify the parameters such as K, $R(t)$, etc., of the problem, when we place ourselves in an approximation of the homogeneous and isotropic universe that leads to the Robertson-Walker $\overline{ds}^2$.

We obtain the two differential equations of Friedman:

$$\frac{2}{R}\frac{d^2R}{dt^2} + \frac{1}{R^2}\left(\frac{dR}{dt}\right)^2 + \frac{Kc^2}{R^2} - \lambda c^2 = \frac{8\pi G}{c^2}\,p \qquad (3)$$

$$\frac{3}{R^2}\left(\frac{dR}{dt}\right)^2 + \frac{3Kc^2}{R^2} - \lambda c^2 = 8\pi G\rho \qquad (4)$$

From these we deduce the equation

$$\frac{d}{dt}(R^3\rho) + \frac{p}{c^2}\frac{d}{dt}(R^3) = 0 \qquad (5)$$

a4. MODELS OF THE UNIVERSE

▪ *Einstein's Model.* In equations (3) and (4), Einstein assumed that $R(t)$ did not change with time and that it was equal to a constant R_0. Moreover, $K = +1$ (a closed universe with a positive curvature) and $p = 0$ (pressure is zero or negligible, or more precisely, p/c^2 is negligible in the presence of ρ). Observation confirms this fact in the observable universe, if one disregards electromagnetic radiation. Thus the first equation gives

$$\lambda = \frac{1}{R_0^2}$$

since the derivatives of R with respect to time are zero, as is p. Bearing in mind the above relationship, the second equation gives

$$\frac{2c^2}{R_0^2} = 8\pi G\rho$$

or

$$R_0 = \frac{c}{2\sqrt{\pi G\rho}}$$

The radius of the universe is inversely proportional to the density of matter; in the limit, it tends toward infinity as density tends toward zero. Observation seems to indicate that the average density of matter in the universe is 10^{-30} grams per cubic centimeter; from this one deduces that R_0 is equal to 10^{28} centimeters, or 10^{23} kilometers.

This model has now been abandoned because it corresponds to a universe without expansion.

▪ *De Sitter's Model.* One assumes that $p = 0$ and $\rho = 0$. This is the limiting case of a universe empty of matter.

Equations (3) and (4) give us

$$\frac{d^2R}{dt^2} - \frac{\lambda c^2}{3} R = 0 \qquad (6)$$

The solution to this differential equation is easily found:

$$R = R_0 \exp\left(\sqrt{\frac{\lambda c^2}{3}}\, t\right)$$

R is an exponential function of t that increases from 0 to $+\infty$ when t varies from $-\infty$ to $+\infty$ if λ is positive, as it is generally assumed to be. If λ is negative, we find that R varies periodically with time; we shall see another model of such a uni-

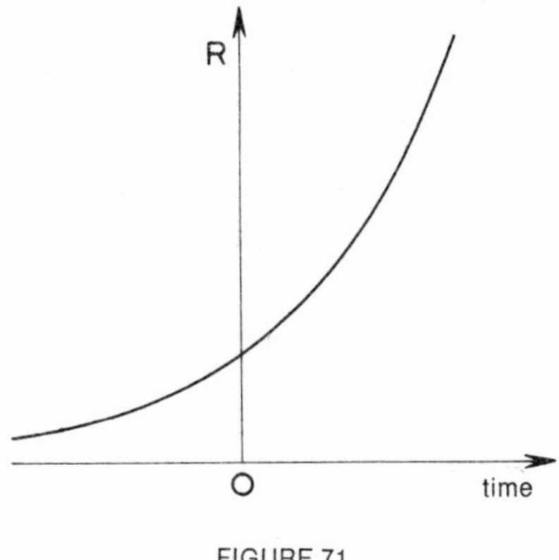

FIGURE 71

verse a little further on. Finally, if λ is zero, we have the universe of Einstein.

Figure 71 represents the variations of R with time when λ is positive.

▪ *Lemaître's Model.* Here it is assumed that $K = +1$ and $p = 0$; these are the hypotheses of Einstein, but the radius R is not constant. This model allows for the expansion of the universe, without also assuming that the universe is empty, as De Sitter did.

In order to integrate the simplified versions of equations (3) and (4) that result from the conditions established, Lemaître as-

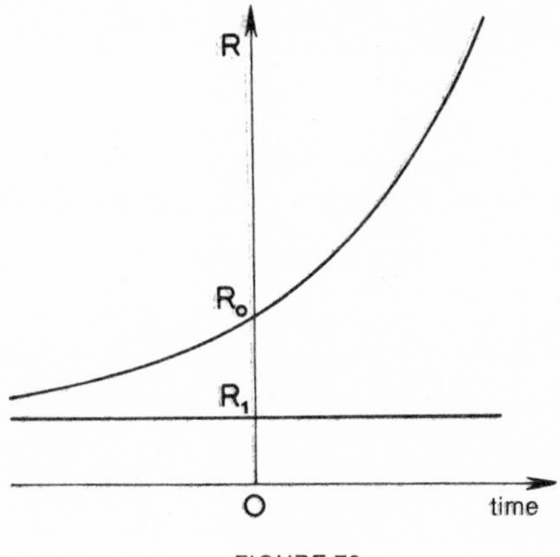

FIGURE 72

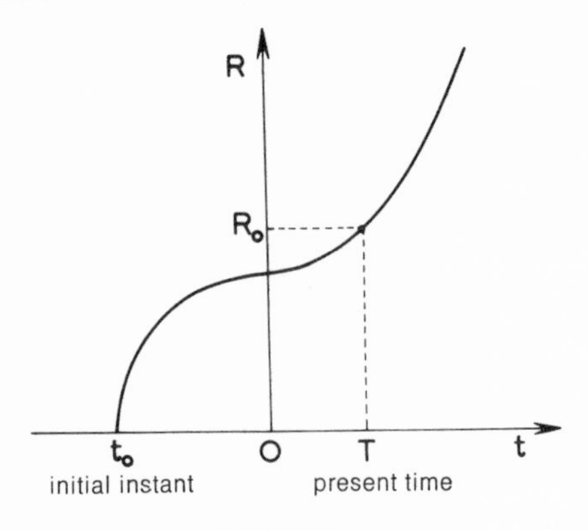

FIGURE 73

sumes that in its initial state the universe was the universe of Einstein. He then finds a variation of R like the one that appears in Figure 72. R increases from R_1 to infinity when t increases from $-\infty$ to $+\infty$, it has the value R_0 when $t = 0$. What is strange about this model is that R should remain equal to a constant R_1 for an infinite period of time. One wonders, in this case, why the radius of the universe would later begin to increase. In fact, Lemaître perfected his model. The universe he obtained later has three periods: an initial explosion during which the radius increases rapidly and a phase of stability of the universe followed, finally, by an expansion (Fig. 73).

▪ *The Einstein–De Sitter Model.* Let us assume that pressure is zero, K is zero, and λ is zero. Friedman's equations become

$$\frac{2}{R}\frac{d^2R}{dt^2} + \frac{1}{R^2}\left(\frac{dR}{dt}\right) = 0$$

$$\frac{3}{R^2}\left(\frac{dR}{dt}\right) = 8\pi G\rho$$

Finally equation (5) shows us that $R^3\rho$ is constant. Let $R^3\rho = A$, A being a constant. Then the second equation above gives us

$$\sqrt{R}(dR) = \sqrt{\frac{8\pi GA}{3}}\,dt$$

which gives us

$$\frac{1}{3}R^{\frac{3}{2}} = \sqrt{\frac{2\pi GA}{3}}\,(t - t_0)$$

t_o being an arbitrary constant. Thus R varies as a ⅔ power of time, if one considers t_0 to be zero, that is, for the time of origin, the time at which R was zero. One can write

$$R = (6\pi GA)^{\frac{1}{3}}t^{\frac{2}{3}}$$

The variation of R as a function of time is shown in Fig. 74. T represents the present time, for which $R = R_o$.

If MN is the tangent to the curve at point M, it is obvious that

$$\frac{PM}{NP} = \frac{dR}{dt} = \frac{R}{NP}$$

for these relationships are equal to the slope of the tangent at M, which is equal to the derivative of function $R(t)$ at point M. But Hubble's law gives us

$$\frac{dR}{dt} = HR$$

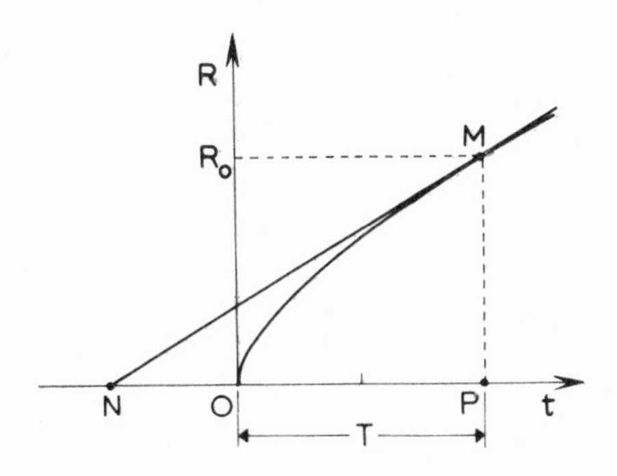

FIGURE 74

Therefore

$$NP = \frac{1}{H}$$

It can also be demonstrated that

$$NP = \frac{3}{2}OP$$

If T is the present time, for which $R = R_o$, we have

$$\frac{1}{H} = \frac{3}{2}T$$

Thus this model relates the "age of the universe," the inverse of Hubble's constant, to the duration of the expansion.

One can also try to integrate Friedman's equations by assuming pressure to be zero as well as λ, but without making any hypothesis about K. It is still true that $R^3\rho = A$, and we then obtain the equation

$$R\left(\frac{dR}{dt}\right)^2 = \frac{8}{3}\pi GA - Kc^2R$$

It is easy to integrate, and we obtain the following result: if $K = +1$, R varies periodically; if $K = -1$, R increases indefinitely (Fig. 75).

In effect, in the first case, the expression $\frac{8}{3}\pi GA - Kc^2R$, which is equal to the product of R times a square number, must remain positive, therefore R must be smaller than $\frac{8}{3}\frac{\pi GA}{Kc^2}$; however, in the second case, the above expression remains positive no matter what the value of R is.

b. THE EVOLUTION OF THE UNIVERSE

b1. RED SHIFT AND EXPANSION. We have attributed the red shift

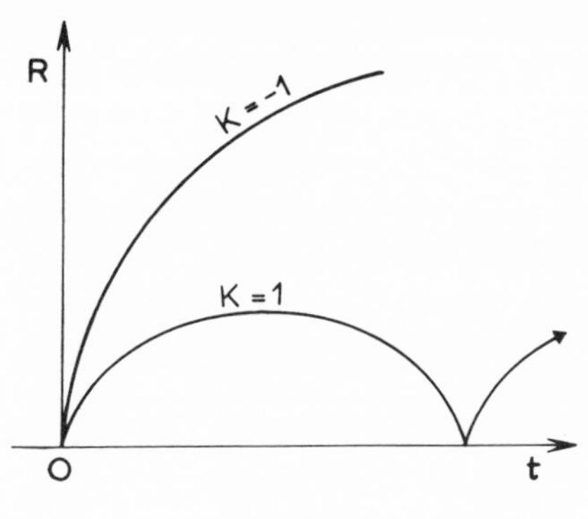

FIGURE 75

of the spectra of remote galaxies to a motion of the galaxies away from the observer. This interpretation is justified by the existence of the Doppler effect, which has been described in its classic form or, when the speeds of the receding motion are very high, in its relativistic form. This apparent flight of the galaxies can be explained very well in a cosmology that allows room for expansion, that is, in those models for which R is an increasing function of time t.

For example, let us consider a photon whose path is radial in relation to the observer; ϕ and θ are zero as well as $d\phi$ and $d\theta$ and, since the photon is describing a path on which ds is zero, the formula that gives the Robertson-Walker $\overline{ds}^2$ is written as follows:

$$\frac{dt}{R(t)} = \frac{dr}{c\left(1 + \frac{Kr^2}{4}\right)}$$

r being a coordinate independent of t. From this we deduce that $dt/R(t)$ is constant (independent of t). It follows that, if a source emits a wavelength at time t' and if the light is received at time t at point origin O, the observer at O will observe a spectral shift. The frequency will be $\lambda + \Delta\lambda$ with:

$$\frac{\Delta\lambda}{\lambda} = \frac{R(t)}{R(t')} - 1$$

We note that $\Delta\lambda/\lambda$ is independent of λ. Moreover, if $R(t)$ increases with t, $R(t)/R(t')$ is greater than 1 and $\Delta\lambda$ is positive (red shift).

Thus we see that in three relativistic models of the universe, the spectral shift is an effect which results immediately from the metrical properties of space-time.

b2. EVOLUTION OF THE GALAXIES. It was once believed that the shape of galaxies depended on their age, and that irregular galaxies were younger than elliptical galaxies. Now, however, it is thought that all galaxies are very nearly the same age. In fact, study of the stars of which they are composed together with what we know about the evolution of the stars suggest ages on the order of 10^{10} years, or close to the inverse of Hubble's constant.

This fact, together with the phenomenon of the expansion of the universe and observation of the cosmological radiation, favors theories like Lemaître's, which attribute a common origin to the universe as a whole and an evolution in a scale of time t (the so-called "big-bang" theory).

Let us consider the universe as a cloud of gas undergoing expansion after the elements have been formed. We must assume that a certain instability is created in certain regions of the cloud, with the result that its homogeneity is locally destroyed. The principal theories accounting for such a phenomenon are those of von Weizsäcker, Hoyle, and Layzer; they more or less came out of James Jeans' theory. In response to local instabilities, portions of the cloud contract, and in so doing, are broken into fragments (Hoyle and Jeans). Layzer, on the other hand, believes that systems of small masses, formed at the beginning of the universe, were later combined due to the force of mutual gravitation.

In any case, the formation of the elements at the origin of the universe must be explained. It is conceivable that one-hundredth of a second after the initial explosion, the universe consisted of a gas of photons, electrons, and neutrinos at a temperature of 100 billion degrees. Protons and neutrons have a density of approximately 100 grams per cubic centimeter. The energy of the photons decreases. The temperature drops to ten billion degrees after ten seconds, and deuterium, He^3, then He^4 begin to form. After a quarter of an hour, the universe contains radiation and matter comprised of 75 percent hydrogen and 25 percent helium, approximately, as well as very small quantities of other elements. It is not until several thousands of years later that electrons unite with the nuclei to form neutral atoms. Several million years later, the galaxies appear.

As for the future evolution of the universe, we have seen that it is very difficult to decide among the different theoretical models which is the right one, given the inadequacy of observation. However, it seems legitimate to try to apply the second principle of thermodynamics to the universe as a whole. This principle governs the way in which energy is transformed. For example, physicists have known since the nineteenth century the equivalence between mechanical work and heat. A certain mechanical operation causes a release of heat which serves to raise the temperature of a body of water, for example, by a certain number of degrees. The second principle tells us that it is impossible to create a mechanical energy by lowering the temperature of a single body of water by so many degrees. In order for a heat engine to function, one must have two sources at different temperatures. The caloric energy travels from the hot source to the cold source and can be tapped and transformed into mechanical energy "en route," although not, in fact, in its totality.

Similarly, in an isolated mass of gas at a certain temperature, a warmer and a cooler region cannot appear. Such a phenomenon, if it occurred, would diminish the value of a certain quantity bound to the system, known as its entropy—which is impossible, according to the second principle. On the contrary, when the temperatures of two sources are gradually equalized, the entropy of the system spontaneously increases, as it should, or remains constant. Entropy therefore seems to be a quantity which characterizes the disorder of a physical system. Such a system, if left to itself, evolves in such a way that its molecules assume, on the average, random positions and velocities, in opposition to what occurs when a certain clearly defined configuration of the system is considered.

So we see that the equivalence between mass and energy discovered by Einstein does not seem to operate in both directions; matter may be transformed into energy, but the opposite does not generally occur.

In any case, if one regards the universe as an isolated system, its entropy can only increase; but the second principle of thermodynamics, as it is expressed in physics, seems to be in contradiction with what we know about the expansion of the universe. This is a mystery which has yet to be solved.

IV. RECENT DEVELOPMENTS

The findings of observation are proliferating rapidly in the field of astrophysics, and contrary to Pascal's famous remark, the imagination of theoreticians in this domain is just as inexhaustible in forming conceptions as nature is in providing material. It is difficult to know *a priori* which of these numerous advances in theory and observation is or will be the most important one for cosmology, for every phenomenon, even one that seems very "local" and particular, may have a direct, though hidden, relationship with the great cosmic processes. Among the numerous new facts and ideas that have recently enriched astrophysics, we have, therefore, retained what seemed to us most important and significant in terms of the present state of knowledge in cosmology, without being absolutely sure that our choice is the best one.

The "black holes" have long been known to theorists as one of the bizarre but seemingly inevitable consequences of the fundamental laws of the theory of relativity, when a sufficiently large mass has been gathered together by gravitation, and when all the known possibilities for resisting collapse have been exhausted. What is new is the idea that, in the absence of direct observation which is, by definition, impossible, the presence of the black holes in the universe might explain certain heretofore grossly conflicting results in the observation of groups of galaxies. It would then be necessary to recognize that it is not simply the refinements of theory, but nature itself that eludes this law of regularity which cosmologists, the modern as well as the ancient, place at the foundation of their systems.

It was almost ten years ago that astronomers discovered quasars, but their strange properties have not yet been explained in a satisfactory way. For a long time the idea prevailed that they are extremely remote, some being at the borders of that part of the universe that can presently be observed; but scientists are no longer so sure about this.

The electrical asymmetry of matter with which we are familiar has long been regarded as an unfathomable mystery. Microphysics has now proved by theory and experiment that this asymmetry (and other asymmetries that have been revealed more recently) does not depend on the laws of nature, which tolerate positive electrons and negative protons as well as "ordinary" electrons and protons, and generally speaking, tolerate "antimatter" as well as matter. Cosmology, which tries to exclude contingent and unexplained findings from its fundamental hypotheses, is, therefore, trying to imagine models of the universe in which matter and antimatter coexist, after being formed in the course of a single process and then separated. So the only contingent circumstance that remains is that we inhabit a region of matter and not of antimatter, a fact to which it would be pointlessly anthropomorphic to try to assign a special explanation.

Although in other areas essential for cosmology, for example, cosmological radiation over ultrashort waves, research is being carried on very actively, no decisive advances have been made of late.

THE BLACK HOLES

In the final stage of its evolution a star may present itself in different states, according to its mass. If this mass is less than 1.6 times the solar mass, the star is evolving toward the state of a white dwarf; its matter is completely ionized and its density is very high (a few hundred tons per cubic centimeter).

If the mass of the star falls between 1.6 and 2 solar masses it will become a star composed of neutrons; the nuclei of the atoms have disintegrated and by another process protons and electrons form neutrons. The density of these stars is enormous and may attain one billion tons per cubic centimeter. The diameter of these stars is of the order of a few dozen kilometers and the law of the conservation of angular momentum decrees that the star rotates at a very high angular velocity (the period of rotation may not exceed a few thousandths of a second). As for the magnetic field, it may reach 10^{12} Gauss (or 10^{8} Teslas by the International System of Measurement). Pulsars have been identified almost certainly as stars composed of neutrons.

Finally, if the mass of the star is greater than two solar masses the pressure increases still further, the density reaches enormous values, and consequently the gravitation on the surface of the star is such that there occur phenomena which only the theory of general relativity can predict.

Indeed, we have seen (IIc3) that light rays are deflected in the vicinity of a mass and that this phenomenon is directly proportional to the intensity of the gravitational field. For a normal star this displacement may reach several seconds of arc, for a white dwarf several minutes, for a star composed of neutrons, several degrees. If we are dealing with an object which is the final stage of a star whose mass is greater than two solar masses, the gravitational field in the vicinity of such an object is so intense that the displacement of light rays may exceed 90°, with the result that the light energy that escapes from the star returns to it. The star is completely invisible to an outside observer, hence the name black hole.

Let us consider Schwarzschild's $\overline{ds}^2$ (IIc4). It contains a constant $2\mu/c^2$, c being the speed of light. In the case of the sun, $2\mu/c^2$ is equal to 2.8 kilometers. We have

$$\overline{ds}^2 = \left(c^2 - \frac{2\mu}{r}\right)\overline{dt}^2 - \frac{\overline{dr}^2}{1 - \dfrac{2\mu}{c^2 r}} - r^2(\overline{d\phi}^2 + \cos^2\phi \times \overline{d\theta}^2)$$

The radius of the star that has been transformed into a black hole is very short (the density of the black hole is very great). Hence it is possible that at a point outside the object, distance r may be less than $2m = 2\mu/c^2$, the coefficient of $\overline{dt}^2$ in the $\overline{ds}^2$ becomes negative, and the coefficient of $\overline{dr}^2$ becomes positive. There is a kind of interchange between space and time accompanied by a singularity for $r = 2m$.

The properties of a black hole are, therefore, altogether unique. Every particle of matter in its vicinity falls into it irresistibly in a finite period of time, if this time is measured according to the scale of time peculiar to the particle, but for an outside observer the particle seems to approach a limiting sphere of radius $R = 2m$, called the Schwarzschild radius, in an infinite period of time (Fig. 76).

In the interior of the black hole Schwarzschild's $\overline{ds}^2$ must be replaced by a $\overline{ds}^2$ calculated for the interior case. It corresponds to a spherical space whose radius of curvature is equal to

$\sqrt{\frac{3/8\,\pi}{\rho}}$, ρ being density.

How can we even conceive of proving the existence of the black holes? They are invisible and reveal their presence only by their mass or their gravitational field. It is possible that a system of double stars might be composed of a normal star and a black hole which would affect the natural motion of the normal star. Moreover, if a black hole were very close, one could observe the fall of particles of matter toward the Schwarzschild sphere. Furthermore, it has been demonstrated that there exists around the black hole an ergosphere (or energy-sphere) which serves as an intermediary between Schwarzschild's sphere and the black hole itself and within which certain exchanges of energy may occur in the form of X rays.

What causes us to suspect the existence of the black holes is that they would permit us to explain the mass defects of groups of galaxies. It is difficult to estimate the mass of a galaxy; however, there do exist various procedures based on relationships that have been demonstrated between mass and certain other physical parameters. In the case of fairly close galaxies one can measure the speed of rotation around the center in relation to the distance to this center by studying the Doppler effect of the rays emitted at different points in the galaxy. The equations of dynamics then provide the distribution of masses within the galaxy and the total mass.

Furthermore, with the so-called virial theorem we can establish the following relationship between V, the average speeds of the particles of a system composed of a large number of bodies M, the total mass of the system, and r, the radius of the system:

$$V^2 = \frac{GM}{2r}$$

(G is the universal constant of gravitation.) If we assume the system to be stable, this relationship enables us to calculate M if we know V and r. Now, the application of this theorem to groups of galaxies gives an average mass for each galaxy in the group which is much higher than those that result from the individual methods referred to above. One may therefore conclude that the missing mass is caused by black holes situated between the galaxies forming the group. It is clear that if the theory of the black holes, which is over forty years old, has regained some favor among astronomers, this is because the current theory of the evolution of stars with large masses leads to it quite naturally.

QUASARS

We know that the very substantial red shift observed in quasars was at first attributed to the expansion of the universe. From this it follows that quasars are extremely remote objects. However, observation seems to show that certain quasars are in the vicinity of relatively close objects to which they clearly appear to be related. Thus, for example, certain explosive galaxies are accompanied by objects which they seem to have ejected and which present the characteristics of quasars. So without being galactic objects, quasars are now thought to be much closer than was originally supposed. But we still have to explain the considerable spectral shift toward red that is observed. The recent hypothesis of Fred Hoyle (*Monthly Notices*, Vol. 155) attributes this phenomenon to the fact that the mass of the proton and the electron is not the same in quasars and in normal galaxies. Indeed, according to the principle of Mach, the mass of a particle depends on the distribution of all the other masses in the universe; new particles formed after an explosion would not have had time to acquire their definitive value, and this is expressed by an obvious red shift which corresponds to the one presented by quasars.

MATTER AND ANTIMATTER IN THE UNIVERSE

After being anticipated by theory, the existence of particles of so-called antimatter has been confirmed by experiment. The first antiparticle to be discovered was the positron, which has the same mass as the electron but a positive electrical charge. If an antielectron encounters an electron, annihilation occurs, and energy is produced in the form of gamma rays composed of two photons emitted in two opposite directions, each of which has an energy of $hv = mc^2$.

Researchers subsequently discovered the antiproton, with a charge equal in magnitude to that of the proton, but negative, as well as the antineutron, with a zero charge, like the neutron, and a mass equal to that of the neutron, but with an opposite magnetic moment.

In certain transformations, when electrons do not appear in pairs, we also find the emission of a neutral particle of very small (probably zero) mass, the neutrino. Experimenters have likewise established the existence of antineutrinos, which can be distinguished from neutrinos only by the nature of a parameter called helicity.

By means of antiparticles, the possibility exists of forming antiatoms; thus, researchers recently obtained the nuclei of certain simple antiatoms.

The properties of antimatter are identical to those of matter; there is a perfect symmetry between the two. In particular, there are no antiphotons, or rather the photons corresponding to antimatter are identical to the photons emitted by matter. Consequently, it is impossible to determine whether remote galaxies are composed of matter or antimatter by studying the electromagnetic radiation they send us. Only observation of the inverse helicity of the neutrinos emitted would permit one to determine this, but such observation is in practice completely impossible.

However, recent theories on the formation of particles at the beginning of the universe suggest the possibility of an equal production of particles and antiparticles, and cosmological models allowing for the two known types of matter are now being considered very seriously.

BIBLIOGRAPHY

BONDI, HERMANN: *Cosmology*, 2nd ed. (Cambridge Monographs on Physics Series). Cambridge: Cambridge University Press; 1960.

COUDERC, P.: *The Expansion of the Universe.* London: Faber & Faber; 1900.

EINSTEIN, ALBERT & INFELD, LEOPOLD: *Evolution of Physics.* New York: Simon and Schuster; 1967.

EVANS, ROBLEY D.: *Atomic Nucleus* (International Series in Pure and Applied Physics). New York: McGraw-Hill; 1955.

FEYNMAN, R. P. *et. al.:* Feynman Lectures on Physics. 3 Vols. Reading, Massachusetts: Addison-Wesley; 1963.

HOYLE, FRED: *Of Men and Galaxies* (John Danz Lecture Series). Seattle: University of Washington Press; 1966.

KOYRE, ALEXANDRE: *From the Closed World to the Infinite Universe.* Baltimore: Johns Hopkins University Press; 1968.

KUHN, THOMAS S.: *Copernican Revolution: Planetary Astronomy in the Development of Western Thought*. Cambridge: Harvard University Press; 1957.

MCVITTLE, G. C.: *Fact and Theory in Cosmology*. New York: Macmillan; 1962.

NORTH, JOHN D.: *Measure of the Universe: A History of Modern Cosmology*. Oxford: Oxford University Press; 1965.

PANNEKOEK, A., TR.: *A History of Astronomy*. Totowa, New Jersey: Rowman & Littlefield; 1961.

SCIAMA, DENNIS W.: *Modern Cosmology*. Cambridge: Cambridge University Press; 1971.

INDEX

A NOTE ON THE TYPE

This book was set on the Linotype in Janson, a recutting made direct from type cast from matrices long thought to have been made by the Dutchman Anton Janson, who was a practicing type founder in Leipzig during the years 1668–87. However, it has been conclusively demonstrated that these types are actually the work of Nicholas Kis (1650–1702), a Hungarian, who most probably learned his trade from the master Dutch type founder Dirk Voskens. The type is an excellent example of the influential and sturdy Dutch types that prevailed in England up to the time William Caslon developed his own incomparable designs from these Dutch faces.

Composed by Kingsport Press, Inc., Kingsport, Tennessee.
Printed and bound by The Book Press, Brattleboro, Vermont.
The book was designed by The Etheredges.